Big Data Imperatives

Enterprise Big Data Warehouse, BI Implementations and Analytics

Soumendra Mohanty
Madhu Jagadeesh
Harsha Srivatsa

Big Data Analytics

ISBN 978-1-4302-4872-9

ISBN 978-1-4302-4873-6 (eBook)

President and Publisher: Paul Manning
Lead Editor: Saswata Mishra
Technical Reviewer: Nitin Sawant
Editorial Board: Steve Anglin, Ewan Buckingham, Gary Cornell, Louise Corrigan,
    Morgan Ertel, Jonathan Gennick, Jonathan Hassell, Robert Hutchinson, Michelle Lowman,
    James Markham, Matthew Moodie, Jeff Olson, Jeffrey Pepper, Douglas Pundick,
    Ben Renow-Clarke, Dominic Shakeshaft, Gwenan Spearing, Steve Weiss, Tom Welsh
Coordinating Editor: Anamika Panchoo
Copy Editor: Michael Sandlin
Compositor: SPi Global
Indexer: SPi Global
Artist: SPi Global
Cover Designer: Anna Ishchenko

Distributed to the book trade worldwide by Springer Science+Business Media New York, 233 Spring Street, 6th Floor, New York, NY 10013. Phone 1-800-SPRINGER, fax (201) 348-4505, e-mail orders-ny@springer-sbm.com, or visit www.springeronline.com. Apress Media, LLC is a California LLC and the sole member (owner) is Springer Science + Business Media Finance Inc (SSBM Finance Inc). SSBM Finance Inc is a Delaware corporation.

For information on translations, please rights@apress.com, or visit www.apress.com.

Apress and friends of ED books may be purchased in bulk for academic, corporate, or promotional use. eBook versions and licenses are also available for most titles. For more information, reference our Special Bulk Sales–eBook Licensing web page at www.apress.com/bulk-sales.

Any source code or other supplementary material referenced by the author in this text is available to readers at www.apress.com/9781430248729. For detailed information about how to locate your book's source code, go to www.apress.com/source-code.

Contents at a Glance

Contents

Preface

The path to here, for us, began in 2011. Data warehouses and BI solutions had become run of the mill; big data was gaining momentum. Sajid Usman (our boss) asked us a very simple but thought-provoking question: "What do you think about big data?" That got us thinking about big data. The definitions are plentiful and situational interpretations are plentiful as well. But a broader set of questions was lurking in our mind. What is the future of traditional data warehousing and BI applications? Are big data solutions a natural evolution of traditional BI applications? Should they co-exist? In our spare time, we started researching this topic, reading published papers, blogs, and other articles. By the end of 2011, a small but unmistakable set of thoughts and ideas began to materialize. It was further enriched by conversations with other practitioners and clients.

This book project began in late 2011. We find ourselves surprised and pleased to still be rolling along with this growing snowball of different thoughts. I (Soumendra) met with Harsha in San Jose during breakfast in a hotel (Marriott San Jose Downtown), we discussed the project and he jumped in to become a co-author. Madhu has been working in the data and analytics area for quite a long time and had always had an inclination to publish; she also agreed to join the group. So, we are only here by accident.

While we are all IT professionals, nobody would mistake us for expert researchers in this area. We are more like museum curators than painters—collecting, organizing, and packaging for wider use the great ideas of an emerging technology area. It turns out that's useful work as well.

After reading a draft, someone recently described the book as certainly a nice collection of thoughts. It was meant as a compliment, and we couldn't agree more. Big data is all about *what we don't know we don't know*, though many of the publications on the subject can look arcane to anyone but the specialist and certainly seem far removed from the reality of applying the techniques. This area is emerging, evolving rapidly, littered with 40+ significant vendor tools and technologies, and most of the technology advancement is coming from open source groups. People like us who make a living by implementing enterprise scale solutions are at a loss and certainly uncomfortable adopting these technologies. But big data is real and is here to stay.

Big Data Imperatives aims to be accessible, to bring forward the interesting nuggets of insight for the enthusiast, and to save the practitioner time in getting work done. We hope it provides you more "a-ha!" moments than "wha...?" moments.

Soumendra Mohanty,
Madhu Jagadeesh, Harsha Srivatsa

About the Authors

Soumendra Mohanty. My interest in big data analytics started during the early part of 2011. At that time, I was struggling to accept the notion that data warehouses and BI solutions were soon to become obsolete. I was more concerned about the fate of thousands of BI practitioners. What should they do? How will they learn this new skill that has all sorts of madness written all over it? Do they need to learn programming skills like Java, Python, NoSQL, etc.? Somewhere along the path, I began to realize the notion of a big data warehouse, hybrid data architectures, and industry use cases that not only need big data solutions but also traditional data warehouses and BI solutions, including analytics. The next thing I knew, I was contributing to articles, whitepapers, and presentations in this space, sharing my thoughts with clients and practitioners.

I am really fortunate to be part of a wonderful and growing community of practitioners and enthusiasts of big data analytics. As more and more companies start adopting big data solutions, I am sure there will be many more interesting aspects of big data that will come to light. I really hope you enjoy reading this book.

Madhu Jagadeesh. I was always passionate about analytics and the various industry analytical applications that we experience in our everyday lives. With the power of big data solutions, analytics has become all the more exciting and path-breaking. This definitely challenged me as a traditional BI and analytics practitioner: to get up to speed on the new advances of technologies and also the ones that are diminishing in this area. While we are learning to work together forming teams of varied niche skills to make big data and analytics implementations, the objective remains the same: achieving business outcomes and working cohesively as a team to achieve these business goals. I feel all industries will plunge into this area; but the pace of adoption would definitely differ based on their current level of maturity and their appetite for taking risks to experiment with new technologies and techniques. To keep our industries competitive it will be a challenge for all of us as practitioners to excel and master this area soon and accelerate our learning and experience to keep pace with the next wave that will hit us! Hope you enjoy reading this book. Happy reading!

Harsha Srivatsa. I consider my work on this book to be a journey of learning, self-discovery, and the realization of a life-long ambition to put my thoughts in print. My career path, which has spanned software engineering, product management, information management consulting, research, and innovation has afforded me the opportunity to work on this project. I have been involved in a number of research and innovation projects involving big data. In addition to having lots of experience with data-related project implementations, I've written extensively on technical subjects. Throughout it all, I've remained fascinated by data and how it speaks to us.

What's fascinating about big data solutions is that they are entirely based on open-source projects and crowd sourcing. A major part of my work comes from developing prototypes using emerging technologies to solve real-world problems; often it is tedious work, as you do not have any other reference points, not even documentation. For this reason, *Big Data Imperatives* not only provides useful explanations of concepts but also guidance regarding the implementation scenarios. I hope this book not only helps the data warehousing and BI practitioners to understand the big data world but also serves as a reference point for all those new to the data and analytics area as well.

About the Technical Reviewer

As Managing Director, Technology, **Nitin Sawant** is the practice lead for technology architecture, BPM, SOA, and cloud at Accenture India. He is an Accenture certified master technology architect (CMTA), leading various initiatives in the emerging technologies of cloud and big data. Nitin has over 17 years of technology experience in developing, designing, and architecting complex enterprise scale systems based on Java, JEE, SOA, and BPM technologies. He received his master's degree in technology in software engineering from the Institute of System Science, National University of Singapore. He graduated with a bachelor's degree in electronics engineering from Bombay University. He is a certified CISSP, CEH, and IBM-certified SOA solutions architect. Nitin has filed three patents in the SOA BPM space and is currently pursuing his PHD in BPM security from BITS Pilani, India.

Acknowledgments

This book wouldn't exist without the efforts of many people. Since this area is emerging and rapidly evolving, no one can claim that they have mastery of the subject; certainly many thoughts in this book are ideas from discussions, publications, blogs, etc. At the end of each chapter we have listed the reference materials we used. The authors gratefully acknowledge some of the many references here, in no particular order.

- The publications in the field of big data, referenced at the end of each chapter.

- The clients and practitioners who have shared their thoughts, problems, and interesting solution ideas with authors.

- Apress, which has invested considerable time and effort in bringing this book to market—particularly Jeffrey Pepper, Saswata Mishra, and Mark Powers have been closely involved in creating these finalized pages.

- The reviewers, who provided valuable feedback during the writing process; and especially we would like to highlight Nitin Sawant's efforts to make the book more relevant to practitioners.

- Everybody who asked questions about big data and the skills of the future needed to succeed, also friends, critics, and well-wishers who supported us through the many hours of writing!

Soumendra

I must convey my sincere gratitude to my loving family (Snigdha, Pratik, and Pratyush) for allowing me to spend hours to write the chapters.

Harsha

I would like to dedicate my work on this book to three important women in my life: my grandmother Indira Ramadurai (1912-1993) who gave me the start and standing in my life; my wife Raji Subramanian for being my pillar of strength and support, and to Illa Dholakia, family friend extraordinaire and purveyor of fine sweetmeats.

I would also like to thank Soumendra and Madhu for the opportunity to collaborate on this book and my Accenture colleagues Umesh Hari, Radhai Sivaraman, Uttama Mukherjee, and Mark Kobe for being most excellent work colleagues.

Madhu

The unconditional support provided by my family made it possible for me to collaborate on this exciting book project. My sincere gratitude goes out to my dear parents, my cute daughter Anusha, and my husband Jagadeesh.

I would like to express my special gratitude and thanks to Soumendra and Harsha for providing me this opportunity to work on this exciting project. It was truly an enriching experience.

My thanks and appreciations also go to my colleagues and friends for their best wishes.

Introduction

You may be wondering—is this book for me? If you are seeking a textbook on Hadoop, then clearly the answer is *no*. This book does not attempt to fully explain the theory and derivation of the various algorithms and techniques behind products such as Hadoop. Some familiarity with Hadoop techniques and related concepts, like NoSQL, is useful in reading this book, but not assumed.

If you are developing, implementing, or managing modern, intelligent applications, then the answer is *yes*. This book provides a practical rather than a theoretical treatment of big data concepts, along with complete examples and recipes for solutions. It develops some insights gleaned by experienced practitioners in the course of demonstrating how big data analytics can be deployed to solve problems.

If you are a researcher in big data, analytics, and related areas, then the answer is *yes*. Chances are, your biggest obstacle is translating new concepts into practice. This book provides a few methodologies, frameworks, and collections of patterns from a practical implementation perspective. This book can serve as a reference explaining how you can leverage traditional data warehousing and BI architectures along with big data technologies like Hadoop to develop big data solutions.

If you are client-facing and always in search of bright ideas to help seize business opportunities, then the answer is *yes*, this book is also for you. Through real-world examples, it will plant ideas about the many ways these techniques can be deployed. It will also help your technical team jump directly to a cost-effective implementation approach that can handle volumes of data previously only realistic for organizations with large technology resources.

Roadmap

This book is broadly divided into three parts, covering concepts and industry-specific use cases, Hadoop and NoSQL technologies, and methodologies and new skills like those of the data scientist.

Part 1 consists of chapters 1 to 3. Chapter 1 introduces big data and its role in the enterprise. This chapter will get you set up for all of the chapters that follow. Chapter 2 covers the need for a new information management paradigm. It explains why the traditional approaches can't handle the big data scale and what you need to do about this. Chapter 3 discusses several industry use cases, bringing to life several interesting implementation scenarios.

Part 2 consists of chapters 4 to 6. Chapter 4 presents the technology evolution, explains the reason for NoSQL data bases, etc. Given that background, Chapter 5 presents application architectures for implementing big data and analytics solutions. Chapter 6 then gives you a first look at NoSQL data modeling techniques in a distributed environment.

Part 3 of the book consists of chapters 7 to 9. Chapter 7 presents a methodology for developing and implementing big data and analytics solutions. Chapter 8 discusses several additional technologies like in-memory data grids and in-memory analytics. Chapter 9 presents the need for a new breed of skills (a.k.a. "data scientist"), shows how it is different from traditional data warehousing and BI skills, tells you what the key characteristics are, and also covers the importance of data visualization techniques.

CHAPTER 1

"Big Data" in the Enterprise

Humans have been generating data for thousands of years. More recently we have seen an amazing progression in the amount of data produced from the advent of mainframes to client server to ERP and now everything digital. For years the overwhelming amount of data produced was deemed useless. But data has always been an integral part of every enterprise, big or small. As the importance and value of data to an enterprise became evident, so did the proliferation of data silos within an enterprise. This data was primarily of structured type, standardized and heavily governed (either through enterprise wide programs or through business functions or IT), the typical volumes of data were in the range of few terabytes and in some cases due to compliance and regulation requirements the volumes expectedly went up several notches higher.

Big data is a combination of transactional data and interactive data. While technologies have mastered the art of managing volumes of transaction data, it is the interactive data that is adding variety and velocity characteristics to the ever-growing data reservoir and subsequently poses significant challenges to enterprises.

Irrespective of how data is managed within an enterprise, if it is leveraged properly, it can deliver immense business values. Figure 1-1 illustrates the value cycle of data, from raw data to decision making. In the early 2000s, the acceptance of concepts like Enterprise Data Warehouse (EDW), Business Intelligence (BI) and analytics, helped enterprises to transform raw data collections into actionable wisdom. Analytics applications such as customer analytics, financial analytics, risk analytics, product analytics, health-care analytics became an integral part of the business applications architecture of any enterprise. But all of these applications were dealing with only one type of data: structured data.

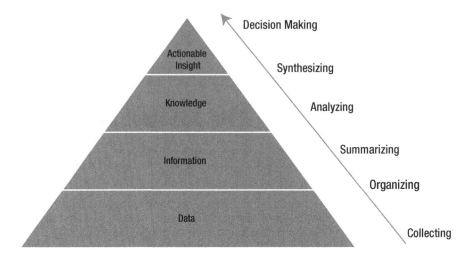

Figure 1-1. *Transforming raw data into action-guiding wisdom*

The ubiquity of the Internet has dramatically changed the way enterprises function. Essentially most every business became a "digital" business. The result was a data explosion. New application paradigms such as web 2.0, social media applications, cloud computing, and software-as-a-service applications further contributed to the data explosion. These new application paradigms added several new dimensions to the very definition of data. Data sources for an enterprise were no longer confined to data stores within the corporate firewalls but also to what is available outside the firewalls. Companies such as LinkedIn, Facebook, Twitter, and Netflix took advantage of these newer data sources to launch innovative product offerings to millions of end users; a new business paradigm of "consumerism" was born.

Data regardless of type, location, and source increasingly has become a core business asset for an enterprise and is now categorized as belonging to two camps: *internal data* (enterprise application data) and *external data* (e.g., web data). With that, a new term has emerged: *big data*. So, what is the definition of this all-encompassing arena called "big data"?

To start with, the definition of big data veers into 3Vs (exploding data volumes, data getting generated at high velocity and data now offering more variety); however, if you scan the Internet for a definition of big data, you will find many more interpretations. There are also other interesting observations around big data: it is not only the 3Vs that need to be considered, rather when the scale of data poses real challenges to the traditional data management principles, it can then be considered a big data problem. The heterogeneous nature of big data across multiple platforms and business functions makes it difficult to be managed by following the traditional data management principles, and there is no single platform or solution that has answers to all the questions related to big data. On the other hand, there is still a vast trove of data within the enterprise firewalls that is unused (or underused) because it has historically been too voluminous and/or raw (i.e., minimally structured) to be exploited by conventional information systems, or too costly or complex to integrate and exploit.

Big data is more a concept than a precise term. Some categorize big data as a volume issue, only to petabyte-scale data collections (> one million GB); some associate big data with the variety of data types even if the volume is in terabytes. These interpretations have made big data issues situational.

The pervasiveness of the Internet has pushed generation and usage of data to unprecedented levels. This aspect of digitization has taken a new meaning. The term "data" is now expanding to cover events captured and stored in the form of text, numbers, graphics, video, images, sound, and signals.

Table 1-1 illustrates the measures of scale of data.

Table 1-1. *Measuring Big Data*

1000 Gigabytes (GB) = 1 Terabyte (TB)
1000 Terabytes = 1 Petabyte (PB)
1000 Petabytes = 1 Exabyte (EB)
1000 Exabytes = 1 Zettabyte (ZB)
1000 Zettabytes = 1 Yottabyte (YB)

Is big data a new problem for enterprises? Not necessarily.

Big data has been of concern in few selected industries and scenarios for some time: physical sciences (meteorology, physics), life sciences (genomics, biomedical research), financial institutions (banking, insurance, and capital markets) and government (defense, treasury). For these industries, big data was primarily a data volume problem, and to solve these data-volume-related issues they had heavily relied on a mash-up of custom-developed technologies and a set of complex programs to collect and manage the data. But, when doing so, these industries and vendor products generally made the total cost of ownership (TCO) of the IT infrastructure rise exponentially every year.

CIOs and CTOs have always grappled with dilemmas like how to lower IT costs to manage the ever-increasing volumes of data, how to build systems that are scalable, how to address performance-related concerns to meet business requirements that are becoming increasingly global in scope and reach, how to manage data security, and privacy and data-quality-related concerns. The polystructured nature of big data has made the concerns increase in manifold ways: how does an industry effectively utilize the poly-structured nature of data (structured data like database content, semi-structured data like log files or XML files and unstructured content like text documents or web pages or graphics) in a cost effective manner?

We have come a long way from the first mainframe era. Over the last few years, technologies have evolved, and now we have solutions that can address some or all of these concerns. Indeed a second mainframe wave is upon us to capture, analyze, classify, and utilize the massive amount of data that can now be collected. There are many instances where organizations, embracing new methodologies and technologies, effectively leverage these poly-structured data reservoirs to innovate. Some of these innovations are described below:

- Search at scale

- Multimedia content

- Sentiment analysis

- Enriching and contextualizing data

- Data discovery or exploratory analytics

- Operational analytics or embedded analytics

In this chapter, we will briefly discuss these use cases; there are several more such use cases, which will be discussed in later chapters.

Search at Scale

In the early days of the Internet, search was primarily used to page through simple lists of results, matching the search objective or key words. Search as a technology has evolved immensely since then. Concepts like iteratively refining a search request by selecting (or excluding) clusters or categories of results, parametric search and guided navigation, type-ahead query suggestions, auto-spelling correction and fuzzy matching (matching via synonyms, phonetics, and approximate spelling) have revolutionized effective means of searching and navigating large volumes of information.

Using natural language processing (NLP) technologies and semantic analysis, it is possible to automatically classify and categorize even big-data-size collections of unstructured content; web search engines like Google, Yahoo!, and Bing are exploiting these advances in technologies today.

Multimedia Content

Multimedia content is fascinating, as it consists of user-generated content like photos, audio files, and videos. From a user perspective this content contains a lot of information: e.g., where was the photo taken, when it was taken, what was the occasion, etc. But from a technology perspective all this metadata needs to be manually tagged with the content to make some meaning out of it, which is a daunting task. Analyzing and categorizing images is an area of intense research. Exploiting this type of content at big data scale is a real challenge. Recent technologies like automatic speech-to-text transcription and object-recognition processing (Content-Based Image Retrieval, or CBIR) are enabling us to structure this content in an automated fashion. If these technologies are used in an industrialized fashion, significant impacts could be made in areas like medicine, media, publishing, environmental science, forensics, and digital asset management.

Sentiment Analysis

Sentiment analysis technology is used to automatically discover, extract, and summarize the context behind unstructured content. It helps in discovering sentiments and opinions and polarity analysis concerning everything from ideas and issues to people, products, and companies. The most cited use case of sentiment analysis is *brand* or *reputation analysis*. The task entails collecting data from select web sources (industry sites, the media, blogs, forums, social networks, etc.), cross-referencing this content with target entities represented in internal systems (services, products, people, programs, etc.), and extracting and summarizing the sentiments expressed in this cross-referenced content.

Companies have started leveraging sentiment analysis technology to understand the voice of consumers and take timely actions such as the ones specified below:

- Monitoring and managing public perceptions of an issue, brand, organization, etc. (called *reputation monitoring*)

- Analyzing reception of a new or revamped service or product

- Anticipating and responding to potential quality, pricing, or compliance issues

- Identifying nascent market growth opportunities and trends in customer demand

Enriching and Contextualizing Data

While it is a common understanding that there is a lot of noise in unstructured data, once you are able to collect, analyze, and organize unstructured data, you can then potentially use it to merge and cross-reference with your enterprise data to further enhance and contextualize your existing structured data. There are already several examples of such initiatives across companies where they have extracted information from high-volume sources like chat, website logs, and social networks to enrich customer profiles in a Customer Relationship Management (CRM) system. Using innovative approaches like Facebook ID and Google ID, several companies have started to capture more details of customers, thereby improving the quality of master data management.

Data Discovery or Exploratory Analytics

Data discovery or exploratory analytics is the process of analyzing data to discover something that had not been previously noticed. It is a type of analytics that requires an open mind and a healthy sense of curiosity to delve deep into data: the paths followed during analysis are in no pre-determined patterns, and success is heavily dependent on the analyst's curiosity as they uncover one intriguing fact and then another, till they arrive at a final conclusion.

This process is in stark contrast to conventional analytics and Online Analytical Processing (OLAP) analysis. In classic OLAP, the questions are pre-defined with additional options to further drill down or drill across to get to the details of the data, but these activities are still confined to finite sets of data and finite sets of questions. Since the activity is primarily to confirm or refute hypotheses, classic OLAP is also sometimes referred to as Confirmatory Data Analysis (CDA).

It is not uncommon for analysts cross-referencing individual and disconnected collections of data sets during the exploratory analysis activity. For example, analysts at Walmart cross-referenced big data collections of weather and sales data and discovered that hurricane warnings trigger sales of not just flashlights and batteries (expected) but also strawberry Pop Tarts breakfast pastries (not expected). And they also found that the top-selling pre-hurricane item is beer (surprise again).

It is interesting to note that Walmart chanced upon this discovery not due to the result of exploratory analytics (as is often reported), but due to conventional analytics.

In 2004, with hurricane Frances approaching, Walmart analysts analyzed their sales data from their data warehouse; they were looking for any tell-tale signs of sales that happened due to the recently passed hurricane Charley. They found beer and pastries were the most-purchased items in a pre-hurricane timeframe, and they took action to increase supplies of these products stores in Frances's path.

The fascinating aspect of Walmart's example is imagining what could happen if we leverage machine-learning algorithms to discover such correlations in an automated way.

Operational Analytics or Embedded Analytics

While exploratory analytics are for discovery and strategies, operational analytics are to deliver actionable intelligence on meaningful operational metrics in real or near-real time. The realm of operational analytics is in the machine-generated data and machine-to-machine interaction data. Companies (particularly in sectors like telecommunications, logistics, transport, retailing, and manufacturing) are producing real-time operational reporting and analytics based on such data and significantly improving agility, operational visibility, and day-to-day decision making as a result.

Dr. Carolyn McGregor of the University of Ontario is using big data and analytics technology to collect and analyze real-time streams of data like respiration, heart rate, and blood pressure readings captured by medical equipment (with electrocardiograms alone generating 1,000 readings per second) for early detection of potentially fatal infections in premature babies.

Another fascinating example is in the home appliances area. Fridges can be embedded with analytics modules that sense data from the various items kept in the fridge. These modules give readings on things like expiry dates and calories and provides timely alerts either to discard or avoid consuming the items.

Realizing Opportunities from Big Data

Big data is now more than a marketing term. Across industries, organizations are assessing ways and means to make better business decisions utilizing such untapped and plentiful information. That means as the big-data technologies evolve and more and more business use cases come into the fray, the need for groundbreaking new approaches to computing, both in hardware and software, are needed.

As enterprises look to innovate at a faster pace, launching innovative products and improve customer services, they need to find better ways of managing and utilizing data both within the internal and external firewalls. Organizations are realizing the need for and the importance of scaling up their existing data management practices and adopting newer information management paradigms to combat the perceived risk of reduced business insight (while the volume of data is increasing rapidly, it is also posing an interesting problem). So an organization's ability to analyze that data to find meaningful insights is becoming increasingly complex.

This is why analyst group IDC defines the type of technology needed to tackle big data as: "A new generation of technologies and architectures, designed to economically extract value from very large volumes of a wide variety of data, by enabling high-velocity capture, discovery, and/or analysis."

Big data technology and capability adoption across different enterprises is varied, ranging from web 2.0 companies such as Google, LinkedIn, and Facebook (their business being wholly dependent on these technologies) to Fortune 500 companies embarking on pilot projects to evaluate how big data capability can co-exist with existing traditional data management infrastructures. Many of the current success stories with big data have come about with companies enabling analytic innovation and creating data services, embedding a culture of innovation to create and propagate new database solutions, enhancing existing solutions for data mining, implementing predictive analytics, and machine learning techniques, complemented by the creation of new skills and roles such as data scientists, big data architects, data visualization specialists, and data engineers leveraging NoSQL products, among others. These enterprises' experiences in the big data landscape are characterized by the following categories: innovation, acceleration, and collaboration.

Innovation

Innovation is characterized by the usage of commodity hardware and distributed processing, scalability through cloud computing and virtualization, and the impetus to deploy NoSQL technologies as an alternative to relational databases. Open-source solution offerings from Apache such as the Hadoop ecosystem are getting into mainstream data management, with solution offerings from established companies such as IBM, Oracle, and EMC, as well as upcoming startups such as Cloudera, HortonWorks, and MapR. The development of big data platforms is perhaps the logical evolution of this trend, resulting in a comprehensive solution across the access, integration, storage, processing, and computing layers. Enterprises will continue to establish big data management capabilities to scale utilization of these innovative offerings, realizing growth in a cost- effective manner.

Acceleration

Enterprises across all industry domains are beginning to embrace the potential of big data impacting core business processes. Upstream oil and gas companies collect and process sensor data to drive real-time production operations, maintenance, and reliability programs. Electronic health records, home health monitoring, tele-health, and new medical imaging devices are driving a data deluge in a connected health world. Emerging location-based data, group purchasing, and online leads allow retailers to continuously listen, engage, and act on customer intent across the purchasing cycle. Mobile usage data for telecom service providers unlock new business models and revenue streams from outdoor ad placements.

The imperative for these enterprises is to assess their current Enterprise Information Management (EIM) capabilities, adopt and integrate big data initiatives and embark on programs to enhance their business capabilities and increased competitiveness.

Collaboration

Collaboration is the new trend in the big data scenario, whereby data assets are commoditized, shared, and offered as a product of data services. Data democratization is a leading motivator for this trend. Large data sets from academia, government, and even space research are now available for the public to view, consume, and utilize in creative ways. Data.gov is an example of a public service initiative where public data is shared and has sparked similar initiatives across the globe. Big data use cases are reported in climate modeling, political campaign strategy, poll predictions, environment management, genetic engineering, space science, and other areas.

Data aggregators, data exchanges and data markets such as those from InfoChimps, Factual, Microsoft Azure market place, Axciom and others have come up with data service offerings whereby "trusted" data sets are made available for free or on a subscription basis. This is an example where data sets are assessed with an inherent value as data products.

Crowdsourcing is a rapidly growing trend where skilled and passionate people collaborate to develop innovative approaches to develop insights and recommendation schemes. Kaggle offers a big data platform for predictive modeling and analytic competitions effectively making "data science a sport." Visual.ly offers one of the largest data visualization showcases in the world, effectively exemplifying the collective talent and creativity of a large user base.

The possibilities for new ideas and offerings will be forthcoming at a tremendous rate in the coming years. As big data technologies mature and become easier to deploy and use, expect to see more solutions coming out especially merging with the other areas of cloud, mobile, and social media.

There is widespread awareness of the revenue and growth potential from enterprise data assets. Data management is no longer seen as a cost center. Enterprise information management is now perceived to be a critical initiative that can potentially impact the bottom line. Data-driven companies can offer services like data democratization and data monetization to launch new business models.

■ **Note**   Data democratization, the sharing of data and making data available to anyone that was once available only to a select few, is leading to creative usage of data such as data mashups and enhanced data visualization. Data monetization (i.e., the business model of offering data sets as a shareable commodity) has resulted in data service providers such as data aggregators and data exchanges.

Big data analytics can thus enable new business opportunities from an operational perspective. They provide effective utilization of data assets and rapid data insights into business processes and enterprise applications and also enhanced analytical capabilities to derive deeper meaningful insights in a rapid fashion, action on business strategies through these enhanced insights into the business and exploitation of missed opportunities in areas previously overlooked. These opportunities arise from the key premise in big data: all data has potential value if it can be collected, analyzed, and used to generate actionable insight and enhance operational business capabilities.

New Business Models

There is a growing awareness and realization that big data analytics platforms are enabling new business models that were previously not possible or were difficult to realize. Utilizing big data technologies and processes holds the promise for improving operational efficiencies and generation of more revenues from new and/or enhanced sales channels.

Enterprises have already realized the benefits obtained by managing enterprise data as an integral and core asset to manage their business and gain competitive advantage from enhanced data utilization and insight.

Over the years, tremendous volumes of data have been generated. Many enterprises have had the foresight not to discard these data and headed down the path to establish enhanced analytical capabilities by leveraging large-scale transactional, interaction data and lately social media data and machine-generated data. Even then, Forrester estimates that only 1 to 1.5 percent of the available data is leveraged. Hence, there is the tantalizing picture of all the business opportunities that can come about with increased utilization of available data assets and newer ways of putting data to good use.

New Revenue Growth Opportunities

The big data age has enabled enterprises of all sizes ranging from startups to small business and established large enterprises to utilize a new generation of processes and technologies. In many instances the promise of overcoming the scalability and agility challenges of traditional data management, coupled with the creative usage of data from multiple sources, have enterprise stakeholders taking serious notice of their big data potential.

McKinsey's analysis (summarized in Figure 1-2) indicates that big data has the potential to add value across all industry segments. Companies likely to get the most out of big data analytics include:

- Financial services: Capital markets generate large quantities of stock market and banking transaction data that can help in fraud detection, maximizing successful trades, etc.

- Supply chain, logistics, and manufacturing: With RFID sensors, handheld scanners, and on-board GPS vehicle and shipment tracking, logistics and manufacturing operations produce vast quantities of information to aid in route optimization, cost savings, and operational efficiency.

- Online services and web analytics: Firms can greatly benefit from increasing their customer intelligence and using it for effective cross-selling/up.

- Energy and utilities: "Smart grids" and electronic sensors attached to machinery, oil pipelines and equipment generate streams of incoming data that can be used for preventive means to avoid disastrous failures.

- Media and telecommunications: Streaming media, smartphones, tablets, browsing behavior and text messages aid in analyzing the user interests and behavior and improve customer retention and avoid churn.

- Health care and life sciences: Analyzing electronic medical records systems in aiding optimum patient treatment options and analyzing data for clinical studies can heavily influence both individual patients' care and public health management and policy.

- Retail and consumer products: Retailers can analyze vast quantities of sales transaction data and understand the buying behaviors, as well as make effective individual-focused customized campaigns by analyzing social networking data.

	Volume of Data	Velocity of Data	Variety of Data	Under-Utilized Data ('Dark Data')	Big Data Value Potential
Banking and Securities	High	High	Low	Medium	High
Communications & Media Services	High	High	High	Medium	High
Education	Very Low	Very Low	Very Low	High	Medium
Government	High	Medium	High	High	High
Healthcare Providers	Medium	High	Medium	Medium	High
Insurance	Medium	Medium	Medium	Medium	Medium
Manufacturing	High	High	High	High	High
Chemicals & Natural Resources	High	High	High	High	Medium
Retail	High	High	High	Low	High
Transportation	Medium	Medium	Medium	High	Medium
Utilities	Medium	Medium	Medium	Medium	Medium

Figure 1-2. Big data value across industries

When big data is distilled and analyzed in combination with traditional enterprise data, enterprises can develop a more thorough and insightful understanding of their business, which can lead to enhanced productivity, a stronger competitive position, and greater innovation—all of which can have a significant impact on the bottom line.

For example, collecting sensor data through in-home health-care monitoring devices can help analyze patients' health and vital statistics proactively. This is especially critical in case of elderly patients. Health-care companies and medical insurance companies can then make time interventions to save lives or prevent expenses by reducing hospital admissions costs.

The proliferation of smart phones and other GPS devices offers advertisers an opportunity to target consumers when they are in close proximity to a store, a coffee shop, or a restaurant. This opens up new revenue for service providers and offers many businesses a chance to target new customers.

Retailers usually know who buys their products. Use of social media networks and web-log files from their e-commerce sites can help them understand who didn't buy and why they chose not to. This can enable much more effective micro customer segmentation and targeted marketing campaigns, as well as improve supply chain efficiencies.

Companies can now use sophisticated metrics to better understand their customers. To better manage and analyze customer information, companies can create a single source for all customer interactions and transactions. Forrester believes that organizations can maximize the value of social technologies by taking a 720-degree view of their customers instead of the previous 360-degree view. In the telecom industry, applying predictive models to manage customer churn has long been known as a significant innovation; however, today the telecom companies are exploring new data sources like customers' social profiles to further understand customer behavior and perform micro-segmentations of their customer base. Companies must manage and analyze their customers' profiles to better understand their interactions with their networks of friends, family, peers, and partners. For example, using social relationships the company can further analyze whether customer attrition from their customer base is also influencing similar behavior from a host of other customers who have social connections with the same customer. By doing this kind of linkage analysis companies can better target their retention campaigns and increase their revenue and profit.

■ **Note**   What the "720-degree customer view" involves is compiling a more comprehensive (some might say "intrusive") portrait of the customers. In addition to the traditional 360-degree view of the customer's external behavior with the world (i.e., their buying, consuming, influencing, churning, and other observable behaviors), you add an extra 360 degrees of internal behavior (i.e, their experiences, propensities, sentiments, attitudes, etc.) culled from behavioral data sources and/or inferred through sophisticated analytics. (Source: *Targeted Marketing: When Does Cool Cross Over to Creepy?* James Kobielus October 30, 2012.)

Taming the "Big Data"

Big data promises to be transformative. With technology advances, companies now have access to effectively deal with large amounts of data and data from various sources. If this data is put to effective usage, companies can deliver substantial top- and bottom-line benefits. Figure 1-3 provides an illustration of how the evolution of big data happened over different timelines.

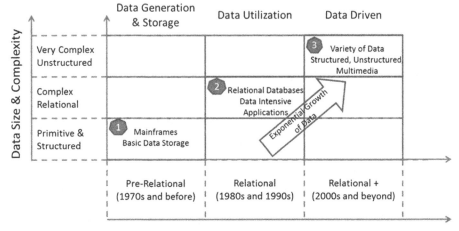

Figure 1-3. *The evolution of big data*

Another key aspect of leveraging big data is to also understand where it can be used, when it can be used, and how it can be used. Figure 1-4 is an illustration of how the value drivers of big data are aligned to an organization's strategic objectives.

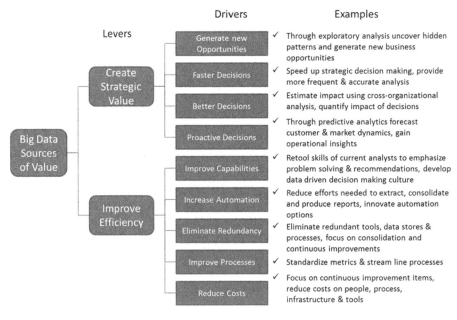

Figure 1-4. *The value drivers of big data*

In some industries big data has spurred entirely new business models. For example, retail banking has started to exploit social media data to create tailored products and offerings for customers in capital markets; due to the onset of algorithmic trading, massive amounts of market data are getting captured, which in turn is helping the regulators to spot market manipulation activities in real time. In the retail sector, big data is expediting analysis of in-store purchasing behaviors, customer footprint analysis, inventory optimization, store layout arrangement—all in near-real time.

While every industry uses different approaches and focuses on different aspects from marketing to supply chain, almost all are immersed in a transformation that leverages analytics and big data (see Figure 1-5).

Retail		Manufacturing	
✓ Customer Relationship Management ✓ Store Location & Layout	✓ Fraud Detection & Prevention ✓ Supply-Chain optimization ✓ Dynamic Pricing	✓ Product Research ✓ Engineering Analysis ✓ Predictive Maintenance	✓ Process & Quality Metrics ✓ Distribution Optimization
Financial Services		**Media & Telecommunications**	
✓ Algorithmic Trading ✓ Risk Analysis	✓ Fraud Detection ✓ Portfolio Analysis	✓ Network Optimization ✓ Customer Scoring	✓ Churn Prevention ✓ Fraud Prevention
Advertising & Public Relations		**Energy**	
✓ Demand Signaling ✓ Targeted Advertising	✓ Sentiment Analysis ✓ Customer Acquisition	✓ Smart Grid ✓ Exploration	✓ Operational Modeling ✓ Power-Line Sensors
Government		**Healthcare & Life Sciences**	
✓ Market Governance ✓ Weapon Systems & Counter Terrorism	✓ Econometrics ✓ Health Informatics	✓ Pharmacogenomics ✓ Bioinformatics	✓ Pharmaceutical Research ✓ Clinical Outcomes Research

Figure 1-5. *Industry use cases for big data*

Yet few organizations have fully grasped what big data is and what it can mean for the future. At present most of the big data initiatives are at an experimental stage. While we believe no organization should miss the opportunities that big data offers, the hardest part is knowing how to get started. Before you embark on a big data initiative, you should get answers to the following four questions to help you on your transformation journey:

- Where will big data and analytics create advantages for the company?

- How should you organize to capture the benefits of big data and analytics?

- What technology investments can enable the analytics capabilities?

- How do you get started on the big data journey?

Where Will Big Data and Analytics Create Advantages for the Company?

Understanding where big data can drive competitive advantage is essential to realizing its value. There are quite a number of use cases, but some important ones are customer intimacy, product innovation, and operations efficiency.

Big data puts the customer at the heart of corporate strategy. Information on social-media platforms such as Facebook is particularly telling, with users sharing nearly 30 billion pieces of content daily. Organizations are collecting customer data from interactive websites, online communities, and government and third-party data markets to enhance and enrich the customer profiles. Making use of advanced analytics tools, organizations are creating data mash-ups by bringing together social-media feeds, weather data, cultural events, and internal data such as customer contact information to develop innovative marketing strategies.

Let's look at few other real-world examples of how big data is helping on customer intimacy. US retailer Macy's is using big data to create customer-centric assortments. Moving beyond the traditional data analysis scenarios involving sell-through rates, out-of-stocks, or price promotions within the merchandising hierarchy, the retailer with the help of big data capabilities is now able to analyze these data points at the product or SKU level at a particular time and location and then generate thousands of scenarios to gauge the probability of selling a particular product at a certain time and place: ultimately optimizing assortments by location, time, and profitability.

Online businesses and e-commerce applications have revolutionized customized offerings in real time. Amazon has been doing this for years by displaying products in a "Customers who bought this item also bought these other items" kind of format. Offline advertising like ad placement and determining the prime time slots and which TV programs will deliver the biggest impact for different customer segments are fully leveraging big data analytics.

Big data was even a factor in the 2012 US Presidential election. The campaign management team collated data from various aspects like polling, fundraising, volunteers, and social media into a central database. Then they were able to assess individual voters' online activities and ascertain whether campaign tactics were producing results. Based on the data analysis, the campaign team developed targeted messaging and communications at individual voter levels which prompted exceptionally high turnout: this was considered one of the critical factors in Obama's re-election.

Product Innovation. Not all big data is new data. There is a wealth of information sitting unused within the corporate data repositories or at least not used effectively. Crowdsourcing and other social product innovation techniques are made possible because of big data. It is now possible to transform hundreds of millions of rich tweets, which is a vast trove of unstructured data, into insights on products and services that resonate with consumers. Data as a service is another innovation that has triggered a number of data- driven companies. For example, compiling and analyzing transaction data between retailers and their suppliers and retailers that own this data, can apply sophisticated analytics to pinpoint process-related inefficiencies and use the insights to improve operations, offer additional services to customers, and even replace third-party organizations that currently provide these services, thus generating entirely new revenue streams.

Some data, once captured, can enable long-established companies to generate revenue and improve their products in new ways. GE is planning a new breed of "connected equipment," including its jet engines, CT scanners, and generators armed with sensors that will send terabytes of data over the Internet back to GE product engineers. The company plans to use that information to make its products more efficient, saving its customers billions of dollars annually and creating a new slice of business for GE.

Finally, imagine the potential big data brings to running experiments—taking a business problem or hypothesis and working with large data sets to model, integrate, analyze, and determine what works and what doesn't, refine the process, and repeat. This activity for online webpages is popularly referred to as A/B testing, Facebook runs thousands of experiments daily with one set of users seeing different features than others; Amazon offers different content and dynamic pricing to various customers and makes adjustments as appropriate.

Operations efficiency: At an operational level, there are a lot of machine- generated data that offer a variety of information-rich interactions, including physical product movements captured through radio frequency identification (RFID) and micro-sensors. Machine-generated data, if captured and analyzed during real time, can provide significant process improvement opportunities across suppliers, manufacturing sites, customers, and can lead to reduced inventory, improved productivity, and lower costs.

For example, in a retail chain scenario, it is quite common to have detailed SKU inventory information to identify overstocks at one store that could be sold in another. However, without a big data and analytics platform, the retail chain is constrained to only identify the top 100 overstocked SKUs. By establishing a big data and analytics platform, the detailed SKU level analysis can be done on the entire data set (several terabytes of operational data) and create a comprehensive model of SKUs across thousands of stores. The chain can then quickly move hundreds of millions of dollars in store overstocks to various other stores, thereby reducing the inventory cost at some stores while increasing sales at other stores and overall net gains for the retail chain.

How Should You Organize to Capture the Benefits of Big Data and Analytics?

Big data platforms provide a scalable, robust, and low-cost option to process large and diverse data sets; however, the key is not in organizing and managing large data sets but to generate insights from the data. This is where specialists such as data scientists come into the picture, interpreting and converting the data and relationships into insights.

Data scientists combine advanced statistical and mathematical knowledge along with business knowledge to contextualize big data. They work closely with business managers, process owners, and IT departments to derive insights that lead to more strategic decisions.

Designing business models: "change management" as an organization process always goes through various levels of maturity; in the case of big data analytics, it's all the more important to understand the current maturity level of the organization and then through a gradual change management process enable the organization to achieve the desired level of maturity. Figure 1-6 outlines three stages of maturity. "Initial Level" provides a historic view of business performance: what happened, where it happened, how many times it happened. In the initial level, most of the analysis is reactive in nature and looks backward into historical data. The analysis performed at this level does not have repeatability and in most cases is ad-hoc in nature; the data management platforms and analyst teams are set up on an as-needed basis. The next level of maturity is "Repeatable and Defined:" at this level, you start looking into unique drivers, root causes, cause-effect analysis as well as performing simulation scenarios like "What-If." At this level, the data management platforms are in place and analysts' teams have a pre-defined role and objectives to support. The next level is "Optimized and Predictive": at this level, you are doing deeper data analysis, performing business modeling and simulations with a goal to predict what will happen.

Analytics Process Maturity	Techniques	
Initial ✓ Proves a static, historical view of business performance ✓ Draws on basic scorecards & static reports	✓ Query & Drill-Down ✓ Ad-Hoc Reporting ✓ Standard Reporting	✓ Where is the problem? ✓ How many? How often? Where? ✓ What happened?
Repeatable and Defined ✓ Creates transparency into past & potential future performance drivers ✓ Uses systems & processes to perform a range of descriptive analysis	✓ Segmentation Analysis ✓ Statistical Analysis ✓ Sensitivity Analysis	✓ What are the unique drivers? ✓ Why is this happening? ✓ What if conditions change?
Optimized and Predictive ✓ Offers dynamic , forward-looking insights with quantified trade-offs ✓ Requires high quality integrated data & complex mathematical modeling capabilities	✓ Optimization ✓ Simulation ✓ Predictive Modeling	✓ What is the best that can happen ✓ What would happen if ...? ✓ What could happen next?

Figure 1-6. *Analytics process maturity*

While the analytics process maturity levels help organizations to identify where they are at present and then gives them a road map to get to the desired higher levels of maturity, another critical component in the transformational journey is the organization model. You can have the best tools installed and the best people in your team, but if you do not have a rightly aligned organizational model, your journey becomes tougher.

There are three types of organization models ("decentralized," "shared services," and "independent"), and each one of these models has its pros and cons (see Figure 1-7).

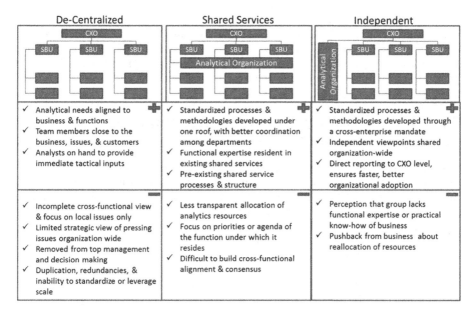

Figure 1-7. *Analytics organization models*

In a "decentralized" model, each business or function will have its own analytics team: for example, sales and marketing will have their own team, finance will have their own team, etc. On the one hand, this enables rapid analysis and execution outcomes, but on the other hand the insights generated are narrow and restrictive to that business function only, and you will not reap the benefit of a broader, game-changing idea. In addition, the focus and drive for analytics is not driven top down from the highest level of sponsorship; as a result, most analytics activities happen in bursts with little to no strategic planning or organizational commitments.

The "shared services" model addresses a few of the shortcomings of the decentralized model by bringing the analytics groups into a centralized model. These "services" were initially governed by bygone systems, existing functions or business units, but with a clear goal to serve the entire organization. While these were standardized processes, the ability to share best practices and organization-wide analytics culture is what makes the shared services model superior to the decentralized model. Insight generation and decision making could easily become a slow process: the reason is that there was no clear owner of this group, and it is quite common to see conflicting requirements, business cases, etc.

The "independent" model is similar to the "shared services" model but exists outside organizational entities or functions. It has direct executive-level reporting and elevates analytics to a vital core competency rather than an enabling capability. Due to the highest level of sponsorship, this group can quickly streamline requirements, assign prioritizations and continue on their insight generation goals.

A centralized analytics unit ensures a broader sweep of insight generation objectives for the entire business. It also addresses another critical area: skills and infrastructure. Many of the roles integral to big data and analytics already exist in most organizations; however, developing a data-driven culture and retaining the rare skills of a data scientist, for instance, are critical to the success of the transformation journey.

What Technology Investments Can Enable the Analytics Capabilities?

Big data and analytics capabilities necessitate transformation of the IT architecture at an appropriate cost. For the last decade or so, organizations have invested millions of dollars in establishing their IT architectures, but for the reasons discussed earlier in this chapter and further influenced by the very changing nature of the data, those investments needs to be critically evaluated. This requires leveraging the old with the new. Unlike the enterprise architecture standards, which are stable and time tested, the big data and analytics architectures are new and still evolving, hence it is all the more important to critically review all the options that exist to make the correct technology investments.

As the complexity of data changes from structured to unstructured, from "clean" in-house data to "noise infected" external data, and from one-dimensional transactional data flow to multi-dimensional interaction data flow, the architecture should be robust and scalable enough to efficiently handle all of these challenges.

At a conceptual level, the big data and analytics technology architecture has five layers, and each layer is specifically designed to handle clear objectives: presentation, application, processing, storage, and integration (see Figure 1-8). The presentation layer provides the functionality to interact with data through process workflow and management. It also acts as a consumption layer through reporting and dashboards and data-visualization tools. The application layer provides mechanisms to apply business logics, transformations, modeling, and other data intensive operations as relevant for business applications and analytics use cases. The processing and storage layers do the heavy-duty process work and store large of volumes of structured and unstructured data in real time or near real time. These layers define the data management and storage criteria consisting of a mix of RDBMS and non-RDBMS technologies. The integration layer acts as a pipe between various enterprise data sources and external data sources; their main job is to help move the desired data and make it available in the storage and processing layer in the big data architecture.

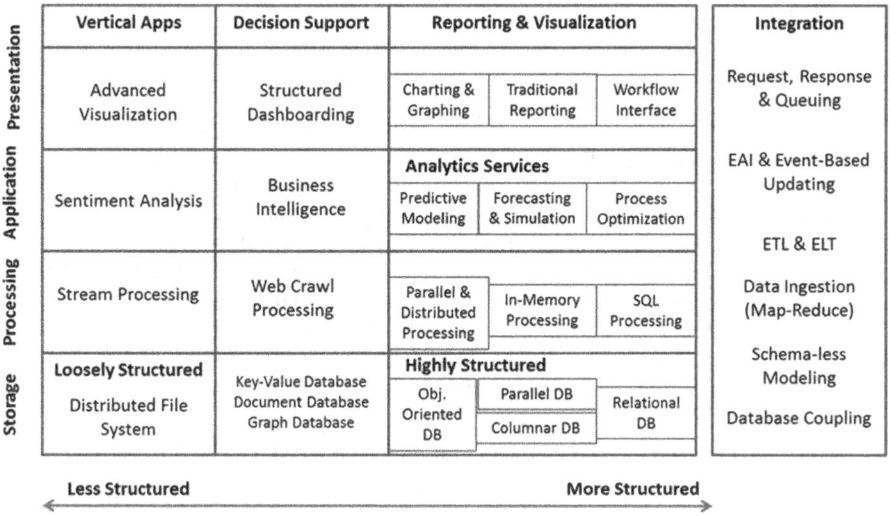

	Vertical Apps	Decision Support	Reporting & Visualization			Integration
Presentation	Advanced Visualization	Structured Dashboarding	Charting & Graphing	Traditional Reporting	Workflow Interface	Request, Response & Queuing
Application	Sentiment Analysis	Business Intelligence	**Analytics Services**			EAI & Event-Based Updating
			Predictive Modeling	Forecasting & Simulation	Process Optimization	ETL & ELT
Processing	Stream Processing	Web Crawl Processing	Parallel & Distributed Processing	In-Memory Processing	SQL Processing	Data Ingestion (Map-Reduce)
Storage	**Loosely Structured**	Key-Value Database Document Database Graph Database	**Highly Structured**			Schema-less Modeling
	Distributed File System		Obj. Oriented DB	Parallel DB Columnar DB	Relational DB	Database Coupling

Less Structured ← → More Structured

Figure 1-8. *Conceptual big data analytics architecture*

Each one of these layers are further grouped to reflect the market segments for new big data and analytics products:

- **Vertical applications,** or product suites, consist of a single vendor providing the entire stack offering. Examples are Hadoop Ecosystem, IBM Big Data Insight, Oracle Exalytics, SAP BI and HANA, among others.

- **Decision support** products specialize in traditional EDW and BI suites.

- **Reporting and visualization tools** are new, and they specialize in how to represent the complex big data and analytics results in an easy-to-understand and intuitive manner.

- **Analytics services** specialize on sophisticated analytics modules, some of them could be cross-functional like claims analytics or customer churn, while some could be very deep in specific areas like fraud detection, warranty analytics, among others.

- **Parallel distributed processing and storage** enable massively parallel processing (MPP), in-memory analytics for more structured data.

- **Loosely structured storage** captures and stores unstructured data.

- **Highly structured storage** captures and stores traditional databases, including their parallel and distributed manifestations.

How Do You Get Started on the Big Data Journey?

For every successful big data implementation, there is an equally successful change management program. To bring the point home, let's discuss the case of a hypothetical traditional big-box retailer. The company had not seen positive same-store sales for years, and the market was getting more competitive. A member of the executive team complained that "online retailers are eating our lunch." Poor economic conditions, changing consumer behaviors, new competitors, more channels, and more data were all having an impact. There was a strong push to move aggressively into e-commerce and online channels. The retailer had spent millions of dollars on one-off projects to fix the problems, but nothing was working. Several factors were turning the company toward competing on analytics: from competitors' investments and a sharp rise in structured and unstructured data to a need for more insightful data.

Transforming analytical capabilities and big data platform begins with a well-thought-out, three-pronged approach (see Figure 1-9).

1. Identify where big data can be a game changer	2. Build future state capability scenarios	3. Define benefits and roadmap
✓ What key business and functional capabilities are required? ✓ What IT capabilities are needed to support and grow the business? ✓ Where are the major gaps in capabilities to support the business?	✓ What are the options for future state business capabilities & technologies? ✓ How do the options compare for capabilities, costs, risks & flexibility? ✓ What functional, analytical, & technology decisions are needed to support these capabilities?	✓ What is the investment's payback period? ✓ What is the implementation roadmap? ✓ What are the key milestones? ✓ What skills are needed? Where are the talent gaps? ✓ What are the risks?

Figure 1-9. *Big data journey roadmap*

Identify where big data can be a game changer. For our big-box retailer, new capabilities were needed if the business had any chance of pulling out of its current malaise and gaining a competitive advantage—the kind that would last despite hits from ever-changing, volatile markets and increased competition. The team engaged all areas of the business, from merchandising, forecasting, and purchasing to distribution, allocation, and transportation, to understand where analytics could improve results. Emphasis was placed on predictive analytics rather than reactive data analysis. So instead of answering why take-and-bake pizza sales are declining, the retailer focused on predicting sales decline and volume shifts in the take-and-bake pizza category over time and across geographic regions. The business also wanted to move from reacting to safety issues to predicting them before they occur. The retailer planned to use social media data to "listen" for problems, which would not only make the company more customer-centric but also provide a shield to future crises. The company planned to set up an analytics organization with four goals in mind:

> Deliver information tailored to meet specific needs across the organization.
>
> Build the skills needed to answer the competition.
>
> Create a collaborative analytical platform across the organization.
>
> Gain a consistent view of what is sold across channels and geographies.

Build future-state capability scenarios. The retailer was eager to develop scenarios for future capabilities, which were evaluated in terms of total costs, risks, and flexibility and determined within the context of the corporate culture. For example, is the business data driven? Or is the company comfortable with hypothesis-based thinking and experimentation? Both are the essence of big data. The company critically reviewed their existing IT architecture in the context of crucial business opportunities, such as leveraging leading-edge technologies and providing a collaboration platform, integrating advanced analytics with existing and new architecture, and building a scalable platform for multiple analytic types. The new technology architecture was finalized to enable the following five key capabilities:

- Predicting customers' purchasing and buying behaviors.

- Developing tailored pricing, space, and assortment at stores.

- Identifying and leveraging elasticity, affinities, and propensities used in pricing.

- Optimizing global data sourcing from multiple locations and business units.

Define benefits and road map. Armed with these capabilities, the next questions revolve around cost-benefit analysis and risks to be mitigated. Does the company have skills in-house or would it be more cost effective to have external resources provide the big data analytics, at least initially? Would it make financial sense to outsource, or should the company persist with internal resources? For each one, do the company and the analytics team have a clear view of the data they need? All these mean significant investment: is there a ROI plan prepared with clear milestones?

The analysts put together a data plan that clearly outlined data needs from acquisition to storage and then to presentation using a self-serve environment across both structured and unstructured data. The systems architecture roadmap was developed consisting of a hybrid Hadoop-based architecture leveraging existing data warehouse platforms. A business road map outlined a multi-million-dollar investment plan that would deliver a positive payback in less than five years.

The company, in its transformation journey, is now positioned to realize four key benefits from its big data and analytics strategy:

- Delivers consistent information faster and more inexpensively.

- Summarizes and distributes information more effectively across the business to better understand performance and opportunities to leverage the global organization.

- Develops repeatable and defined BI and analytics instead of every group reinventing the wheel to answer similar questions.

- Generates value-creating insights yet to be discovered through advanced analytics.

End Points

The massiveness of data and the complex algorithms it requires is an important issue; but it isn't the most important one. To manage big data you don't have to set up a massive scale of hardware infrastructures anymore; cloud services have given us the capability to run very large server clusters at a low startup cost. Open-source projects from Google and Yahoo have created big data platforms such as the Hadoop ecosystem, enabling processing of massive amounts of data in a distributed data-processing paradigm. These technology evolutions have accelerated a new class of data-driven startups, it has reduced both marketing costs and the time it takes for these startups to flourish. And it has allowed startups that were not necessarily data driven to become more analytical as they evolved, such as Facebook, LinkedIn, Twitter, and many others.

Data issues can happen with even less than a terabyte of data. It is not uncommon to see teams of database administrators employed to manage the scalability and performance issues of EDW systems, which are not even on a big data scale as we discussed earlier. The big issue is not that everyone will suddenly operate at petabyte scale; a lot of companies do not have that much data. The more important topics are the specifics of the storage and processing infrastructure and what approaches best suit each problem. How much data do you have, and what are you trying to do with it? Do you need to do offline batch processing of huge amounts of data to compute statistics? Do you need all your data available online to serve queries from a web application or a service API? What is your enterprise information management strategy and how does it co-exist with the big data realm?

References

Snapshot of data activities in an internet minute: Go-Globe.com
MAD Skills: New Analysis Practices for Big Data: VLDB '09, August 24-28, 2009, Lyon, France
The next frontier of innovation, competition and productivity: Mckinsey.com
Bringing Big Data to the Enterprise, IBM, 2012
A Comprehensive List of Big Data Statistics, Wikibon Blog, 1 August 2012
eBay Study: How to Build Trust and Improve the Shopping Experience, KnowIT Information Systems, 8 May 2012

Big Data Meets Big Data Analytics, SAS, 2011

Big Data' Facts and Statistics That Will Shock You, Fathom Digital Marketing, 8 May 2012

IT Innovation Spurs Renewed Growth at www.atkearney.com

Big Data Market Set to Explode This Year, But What is Big Data?, Smart Planet,
    21 February 2012

Corporations Want Obama's Winning Formula, Bloomberg Businessweek,
    21 November 2012

Mapping and Sharing the Consumer Genome, The New York Times, 16 June 2012

GE Tries to Make Its Machines Cool and Connected, Bloomberg Businessweek,
    6 December 2012

GE's Billion-Dollar Bet on Big Data, Bloomberg Businessweek, 26 April 2012

The Science of Big Data at www.atkearney.com

Data Is Useless Without the Skills to Analyze It, Harvard Business Review,
    13 September 2012

MapReduce and MPP: Two Sides of the Big Data Coin, ZDNet, 2 March 2012

Hadoop Could Save You Money Over a Traditional RDBMS, Computerworld UK,
    10 January 2012

eBay Readies Next Generation Search Built with Hadoop and HBase, InfoQ,
    13 November 2011

■ ■ ■

CHAPTER 2

The New Information Management Paradigm

The ubiquitous nature of data and the promises it has shown for enterprises necessitates a new approach to enterprise information management.

What Is Enterprise Information Management?

For an enterprise to carry out its functions, it needs an ecosystem of business applications, data platforms to store and manage the data, and reporting solutions to provide a view into how the enterprise is performing. Large enterprises with multiple strategic business focus areas need many such applications, and as often seen, over the years the enterprise landscape gets into a spaghetti-like situation where it becomes incomprehensible to articulate which application and which data store does what! Various reasons can be attributed to such a state: lack of enterprise-wide data standards, minimal metadata management processes, inadequate data quality and data governance measures, unclear data archival policies and processes, so on.

In order to overcome this problematic situation, enterprise information management as an organization-wide discipline is needed. Enterprise Information Management (EIM) is a set of data management initiatives to manage, monitor, protect, and enhance the information needs of all the stakeholders in the enterprise. In other words, EIM lays down foundational components and appropriate policies to deliver the right data at the right place at the right time to the right users.

Figure 2-1 lists these foundational components and describes the roles they play in the overall business and IT environment of any organization. The goal is management of information, data, and content to meet the needs of the business.

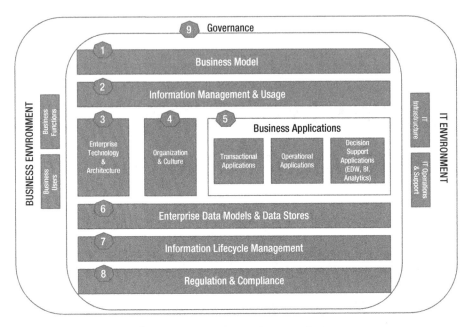

Figure 2-1. *Enterprise information management framework*

The entire framework of EIM has to exist in a collaborative business and IT environment. EIM in a small company or in a startup may not require the same approach and rigor as EIM in a large, highly matured and/or advanced enterprise. The interactions between the components will vary from industry to industry and will be largely governed by business priorities; following a one-size fits all kind of approach to EIM implementation may amount to overkill in many situations. But in general, the following are key components any data-driven enterprise must pay attention to.

1. **Business Model:** This component reflects how your organization operates to accomplish its goals. Are you metrics driven? Are you heavily outsourced, or do you do everything in-house? Do you have a wider eco-system of partners/suppliers or do you transact only with a few? Are your governance controls and accountability measures centralized, decentralized, or federated? The manner in which you get your business objectives successfully implemented down to the lowest levels is your business model.

2. **Information Management and Usage:** A key expectation from an EIM program is to make sure that data and content are managed properly, efficiently, and benefit the business without extra risk.

EIM by definition covers all enterprise information, including reports, forms, catalogs, web pages, databases, and spreadsheets: in short, all enterprise- related structured and unstructured data. All enterprise content may be valuable, and all enterprise content can pose risk. Thus enterprise information should be treated as an asset.

3. **Enterprise Technology and Architecture:** Every enterprise has a defined set of technology and architectures upon which business applications are developed and deployed. Although technology and architecture are largely under the IT department's purview, business requirements and priorities often dictate which technology and architecture to follow. For example, if the company's business is primarily through online applications, then the enterprise technologies and architectures will have a heavy footprint of web-centric technology and architectures. If the company decides they would like to interact with their customers through mobile channels, then you need to make provisions for mobility as well. The choice of technologies and architectures also reflects the type of industry the business belongs to. For example, in the financial services industry where data security and privacy is of utmost concern, it is normal for companies to invest in only a few enterprise-scale platforms, whereas for the retail industry such measures may not be required. So, you will see a plethora of technologies and architectures, including open source systems. The extent to which organizations deploy various technologies and architectures is also a component of EIM.

4. **Organization and Culture:** Who is responsible for managing your data? Is it business or IT or both? If you want your enterprise data to be treated as an asset, you need to define an owner for it. You will need to implement positions and accountabilities for the information being managed. You cannot manage inventory without a manager, and you cannot tackle information management without someone accountable for accuracy and availability.

   EIM helps in establishing a data-driven culture within the enterprise. Roles like data stewards further facilitate the data-driven culture, where right from the CxO levels to the lowest level, people in your organization use data to make informed decisions as opposed to gut-feel decisions.

5. **Business Applications:** How data is used is directly proportional to the value of the data. If you are managing your data as an asset, then the only way to know if that asset has value is to understand how it is used, where it is used, and what impact it is having on the business.

Your transactional applications, operational applications, and decision support applications are all considered to be business applications. You just don't go on creating various types of business applications blindly. The company's business priorities and road maps serve as a critical input to define what kind of business applications need to be built and when. These inputs are then fed into the EIM program to determine what technology and architectures are required, how they will be governed, who will use them, and so on.

6. **Enterprise Data Model and Data Stores:** Enterprise business applications can't run by themselves, so they will need data models and data stores. It is not uncommon to find numerous data models and data stores in an enterprise setup. Too many data models and data stores can cause severe challenges to the enterprise IT infrastructure and make it inefficient; but at the same time, too few data models and data stores will put the company at the risk of running its business optimally. A balance needs to be achieved, and EIM helps in defining policies, standards, and procedures to bring some sanity to the enterprise functioning.

7. **Information Lifecycle Management:** Data and content have a lifecycle. It gets created through transactions and interactions and is used for business-specific purposes; it also gets changed and manipulated following business specific rules, and it gets read and analyzed across the enterprise and then finally reaches a stage where it must be archived for later reference or purged, as it has attained a "use by" state.

EIM defines the data policies and procedures for data usage and thus balances the conflict of retiring data versus the cost and risk of keeping data forever.

Information lifecycle management, if properly defined, also helps in addressing the following common questions:

- What data is needed, and for how long?

- How can my business determine which data is most valuable? Are we sure about the quality of the data in the organization?

- How long should we store this "important" data?

- What are the cost implications of collecting everything and storing it forever? Is it even legal to store data in perpetuity?

- Who is going to go back multiple years and begin conducting new analysis on really old data?

- I don't understand the definitions of data elements, where will I find the metadata information?

There are several important considerations around data quality, metadata management, and master data management that need to be taken into account under the purview of information lifecycle management. A key component of EIM is to establish data lineage (where data came from, who touched it, and where and how it is used) and data traceability (how is it manipulated, who manipulated it, where it is stored, when it should be archived and/or purged).

This function of EIM is extremely valuable for any enterprise. Its absence creates data silos and unmanageable growth of data in the enterprise. In short, you need to know full lineage, definitions, and rules that go with each type of data.

Lack of appropriate data hampering business decision making is an acceptable fact; however, poor data quality leading to bad business decisions is not at all acceptable. Therefore monitoring and controlling the quality of data across the enterprise is of utmost importance. But how do we monitor the quality of data? Using metrics, of course. That means we need a process for defining data quality metrics. Below is a high-level approach to defining DQ metrics your EIM program should follow:

- Define measurable characteristics for data quality. Examples are: state of completeness, validity, consistency, timeliness, and accuracy that make data appropriate for a specific use.

- Monitor the totality of features and characteristics of data that define their ability to satisfy a given purpose.

- Review the processes and technologies involved in ensuring the conformance of data values to business requirements and acceptance criteria.

The end result is a set of measurement processes that associate data quality scores against each business critical data entity. These scores help in quantifying conformance to data quality expectations. Scores that do not meet the specified acceptability thresholds indicate non-conformance.

Closely associated with data quality is the concept of master data management (MDM). MDM comprises a set of processes, governance, policies, standards, and tools that consistently define and manage the master data (i.e. non-transactional data entities) of an organization (which may include reference data).

MDM has the objective of providing processes for collecting, aggregating, matching, consolidating, quality-assuring, persisting, and distributing such data throughout an organization to ensure consistency and control in the ongoing maintenance and application use of this information. A data element, used in various applications, is likely to mean different things in each of them. For example, organizations find it difficult to agree on the definition of very important entities like customer or supplier. At a basic level, MDM seeks to ensure that an organization does not use multiple (potentially inconsistent) versions of the same master data in different parts of its operations, which can occur in large organizations. A common example of poor MDM is the scenario of a bank at which a customer has taken out a mortgage and the bank begins to send mortgage solicitations to that customer, ignoring the fact that the person already has a mortgage account relationship with the bank. This happens because the customer information used by the marketing section within the bank lacks integration with the customer information used by the customer services section of the bank.

Data quality measures provide means to fix data related issues already existing in the organization whereas MDM, if implemented properly, prevents data-quality-related issues from happening in the organization.

Metadata management deals with the softer side of the data-related issues, but it is one of the key enablers within the purview of information lifecycle management. The simplest *definition of metadata is "data about data."* In other words, metadata can be thought of as a label that provides a definition, description, and context for data. Common examples include relational table definitions and flat file layouts. More detailed examples of metadata include conceptual and logical data models.

A famous quote, sometimes referred to as "Segal's Law," states that: "A man with one watch knows what time it is. A man with two watches is never sure." When it comes to the metrics used to make (or explain) critical business decisions, it is not surprising to witness the "we have too many watches" phenomenon as the primary cause of the confusion surrounding the (often conflicting) answers to common business questions, such as:

- How many customers do we have?

- How many products did we sell?

- How much revenue did we generate?

Therefore, another example of metadata is providing clear definitions of what the terms "customers," "products," and "revenue" actually mean.

Metadata is one of the most overlooked aspects of data management, and yet it is the most difficult initiative to implement. Metadata can potentially encompass many levels; from a single data element on the database to a more complex entity, such as customer, for example, which will be a composite of other elements and/or entities.

■ **Note**   The topic of information lifecycle management and especially data quality, master data management and metadata management are itself separate chapters on their own. Here we have given brief overviews about these important concepts as they relate to data and its management.

---

8. **Regulations and Compliance:** Irrespective of which industry your company belongs to, regulatory risk and compliance is of utmost concern. In some industries like financial services and health care, meeting regulatory requirements is of the highest order; whereas other industries may not be exposed to such strict compliance rules. EIM helps you address the regulatory risk that goes with data.

9. **Governance:** Governance is primarily a means to ensure the investments you are making in your business and IT are sustainable. Governance ensures that data standards are perpetuated; data models and data stores are not mushrooming across the enterprise, roles like data stewards are effective, and they resolve conflicts related to data arising within business silos. Most importantly, governance, if enforced in the right spirit, helps manage your data growth and cost impact optimally.

As you can see, there are many components in the EIM framework that must interact with each other in a well-orchestrated manner. When we were discussing EIM, we had mostly discussed data in a generic sense to include all possible types of data and all possible types of data sources (internal data sources as well as external data sources). EIM is at a framework level and does not necessarily anticipate what needs to be done when you are dealing with different kinds of data, especially when we refer to big data characteristics like volume, velocity, and variety.

There are several challenges (some new and some are old, but their impacts are magnified) when we start looking at the finer details of big data and how they impact the EIM framework. Does this mean we will need a radically different approach for the enterprise information management framework?

New Approach to Enterprise Information Management for Big Data

The current approach to EIM has some fundamental challenges when confronted with the scale and characteristics of big data. Below, we will first discuss a few areas related to the very nature of big data and how it is impacting the traditional information management principles.

Type of Data: Traditional information management approaches have focused primarily on structured data. Structured data is stored and managed in data repositories such as relational databases, object databases, network databases, etc. However, today a vast majority of the data being produced is unstructured. By some estimates, about 85 percent to 90 percent of the total data asset is unstructured. This vast amount of unstructured data often goes underutilized because of the complexities involved in the parsing, modeling, and interpretation of the data.

- In the big data scenario, the EIM needs to manage all kinds of data, including traditional structured data, semi-structured, unstructured and poly-structured data, and content such as e-mails, web-page content, video, audio, etc.

Enterprise data modeling: For a long while, data modeling has been an integral part of data management practices, and often you see complex data models developed to store data and manage data in databases. Sometimes this complexity can be attributed to data modeling principles (primarily third normal form design or de-normalized and data-mart-centric design approaches) and sometimes to the inadequacies of the relational database systems. While these data modeling approaches were suitable to managing data at scale and that for structured data only, the big data realm has thrown in additional challenges of *variety* exposing the shortcomings in the technology architecture and the performance of relational databases.

- The cost of scaling and managing infrastructure while delivering a satisfactory consumer experience for newer applications such as web 2.0 and social media applications has proven to be quite steep. This has led to the development of "NoSQL" databases as an alternative technology with features and capabilities that deliver the needs of the particular use case.

Data Integration: For years, traditional data warehousing and data management approaches has been supported by data integration tools for data migration and transportation using Extract-Transform-Load (ETL) approach. These tools run into throughput issues while handling large volumes of data and are not very flexible in handling semi-structured data.

- To overcome these challenges in the big data scenario, there has been a push toward focusing on extract and load approaches (often referred to as *data ingestion*) and applying versatile but programmatically driven parallel transformation techniques such as map-reduce.

Data integration as a process is highly cumbersome and iterative especially when you want to add new data sources. This step often creates delays in incorporating new data sources for analytics, resulting in the loss of value and relevance of the data before it can be utilized. Current approaches to EDW follow the waterfall approach, wherein until you finish one phase, you can't move on to the next phase.

- While this approach has its merits to ensure the right data sources are picked and the right data integration processes are developed to sustain the usefulness of the EDW. In big data scenario, the situation is completely different; one has to ingest a growing number of new data sources, many of them are very loosely defined and probably have no definitions at all, thereby posing significant challenges to the traditional approach of the EDW development lifecycle. In addition, there is a growing need from the business to analyze and get quick insightful and actionable results; they are not ready to wait!

Cost: The costs to manage the data infrastructure (storage, computing, and analysis) have risen significantly due to vendor lock-ins and usage of proprietary technologies. Most enterprises do not even have a clear picture of what kind of data assets they have, where they are located and how much data they have. In many cases, companies do not have a clear enough idea of this asset to predict and anticipate data growth. With all these unknowns, there is a dire need for quicker and more agile approaches to the entire software development lifecycle.

- Now, there are several new technologies and architectures enabling companies with cost effective solutions. We will discuss the SMAQ stack later in this chapter and how it solves the big-data-related issues while at the same time providing a cost effective viable alternative to IT infrastructure.

■ **Note** We are not advising that you sunset all your enterprise IT platforms and adopt the SMAQ stack; but there needs to be a pragmatic approach in developing a big data ecosystem where enterprise platforms and SMAQ systems can co-exist to deliver cost effective solutions for the enterprise. We will discuss these approaches at length in chapters 4, 5, and 6.

Data Quality: There is a debate as to whether data quality principles should be applied to big data scenarios or not. Data quality does have some role to play in big data, as it ensures that the data is well formed, accurate, and can be trusted. Approaching data quality for big data following the traditional route of data profiling (i.e., data cleansing) data monitoring will be extremely difficult; there is too much data to profile, and often you are not so sure about the structure of the data. Moreover, the long time frames for data quality lifecycle (i.e., the approach to remediate data quality issues and deliver "clean" data) does not lend itself too much to agility, which is a key requirement for big data analytics. Data quality issues are more pronounced with transactional data as they are primarily produced due to inadequate checks and controls at the source systems and not so much due to the volume of data.

- Due to these considerations, it is recommended that ongoing data quality initiatives be focused on resolving data quality issues for transactional and reference/master data either closer to the source and/or downstream. For the big data scenarios, there is tremendous value in applying data quality rules to the big data sets and getting an idea of the conformance of such data sets to the applied rules.

MDM: MDM has the inherent goal of reconciling data silos across such categories as customers, products, assets, etc., to produce a consistent, trusted source of critical core business data entities. However, the volume and variety of data in the big data scenarios pose serious challenges to implementing a MDM system for your enterprise.

- The biggest advantage of big data sources (external to the corporate firewalls) is that they help in validating your master entities and in many cases help in enriching them. For example, using Google e-mail ID, Facebook IDs and LinkedIn IDs you can further enrich you customer identification process and improve your conversations with customers through multiple channels.

Metadata Management: Metadata management aims to provide consistent representation and understanding of data definitions across the enterprise. However, due to sheer variety and diversity of data types in big data sets, scaling metadata management to cover big data scenarios becomes very difficult and not economical.

- In many situations, when you are dealing with big data sources, you may not find well-documented definitions associated with data attributes. This is precisely why you should attempt to create a minimum set of documentation consisting of the source, how you accessed it, what access methods (APIs or direct downloads) you applied, what data cleansing methods you applied, what security and privacy measures you applied on the data sets, where you are storing the raw data sets, etc.

Skills: Big data analytics solutions are intended to solve different kinds of problems and they require different kind of skills (data scientist) to accomplish the tasks. The skills like DBA, data integration specialists, and reports development specialists usually are not expected to be competent in collecting, merging, and analyzing data coming from a variety of sources; nor are they expected to have the business acumen to understand the context of the data.

- In big data scenarios, data scientists and data architects rather than database administrators will be in demand to effectively implement the distributed nature of big data processing, ingesting and aggregating data from multiple sources and managing storage, compute, and network resources to handle large data sets.

■ **Note**   In Chapter 9, we will discuss in detail the skills needed to be successful in developing and implementing big data analytics solutions.

In the sections above we discussed what additional considerations need to be put in place under the EIM framework to support big data analytics initiatives in your organization. Big data analytics initiatives are very different in nature. Besides a robust EIM framework, you will need to understand what capabilities need to be put in place to optimally deliver big data analytics initiatives in your organization. What are those?

New capabilities needed for big data

Big data characteristics, especially the velocity and variety aspects of it warrants us to deal with the data and associated events as they happen. We can't afford latency because the data will become useless if you don't act at the time of events happening. In addition, the type of analysis you will make on big data expects it to be much more iterative. The complexity of big data sets also demands better data visualization techniques. Otherwise, it will become tedious and incomprehensible if you follow traditional reporting and dashboard development approaches. In order to move at the speed of business, and maintain competitive advantage, enterprise agility is becoming vital. This means that business requirements need to be developed rather quickly. The organization should have the ability to quickly respond to changing business conditions, and more often than not business will be asking a question, which means data sets are created quickly, analyzed and presented back to business users with possible answers. This further highlights the need for, and the importance of, adoption of agile methods for business intelligence and analytics.

Thus newer capabilities like data discovery, rapid data analysis, advanced data visualization, etc., are needed to effectively handle the big data scenario. We will discuss a few of these capabilities below.

- **Data Discovery:** consists of activities involving locating, cataloging, and setting up access mechanisms for data sources. Such an exercise greatly benefits the enterprise in agile data integration, enriching the content and value of enterprise data assets from both internal and external data sources.

- **Rapid Data Insight:** is the next generation of agile data analysis wherein data from multiple sources can be quickly inspected, cleansed, and transformed with the goal of getting a deeper understanding of the data, spot apparent trends and patterns, and getting an idea of the value of data assets in supporting decision making and analytics. Data insight enables end users to make better "sense" of data assets.

- **Advanced Data Visualization:** is the process whereby reliable data from one or more sources are integrated or mashed up together and visually communicated clearly and effectively through advanced graphical means. This enables you to succinctly present and convey the insight gleaned from large amounts of information and enables better cognitive understanding of such information insight especially to business end users. Another aspect of advanced data visualization is the capability to tell a "data story": i.e., inferences and conclusions that can be articulated using factual data and where a thesis can be developed.

- **Advanced Analytics:** involves the application of business rules, domain knowledge and statistical models, often in-database closer to the data sources themselves, that help in decision making and help answer the questions of "What?" and "Now What?".

- **Data Virtualization:** is a data integration technique that provides complete, high-quality, and actionable information through virtual integration of data across multiple, disparate internal and external data sources. Instead of copying and moving existing source data into physical, integrated data stores (e.g., data warehouses and data marts), data virtualization creates a virtual or logical data store to deliver the data to business users and applications.

- **Data Services:** is described as a modular, reusable, well-defined, business-relevant service that leverages established technology standards to enable the access, integration, and right-time delivery of enterprise data throughout the enterprise and across corporate firewalls. Data services technology provides an abstraction layer between data sources and data consumers.

In a nutshell, adopting an EIM approach addressing all aspects of big data as a platform will enable enterprises to build up their capability progressively and move up in the maturity curve as shown below in Figure 2-2. The maturity model highlights how we see these skills (both technical and business) mapping out in the context of the organizations that have adopted business analytics over time with a view to how this could evolve in the era of big data analytics.

Phase	Old World			New Era
Impact	Pilot	Departmental Analytics	Enterprise Analytics	Big Data Analytics
Skills (IT)	Little or no expertise in Analytics – basic knowledge in BI tools	Data warehouse team focused on performance, availability and data management	Advanced data modelers and data architects key part of the IT department	Analytics Center of Excellence that includes "data scientists"
Skills (Business)	Functional knowledge of BI tools	Few business analysts, limited usage of advanced analytics	Savvy analytical modelers, data stewards and statisticians utilized	Complex problem solving integrated into Analytics Center of Excellence, Deep business analysis knowledge, data exploration and analysis capability
Technology & Tools	Simple historical BI reporting and dashboards	Data warehouse implemented broad usage of BI tools, limited analytics data marts	Analyticsplatforms, Data Visualization platforms, limited usage of parallel processing and analytical appliances/sandboxes	Widespread adoption of analytics sandboxes, appliances for multiple workloads, Architecture and governance for emerging technologies
Financial Impact	No substantial financial impact. No ROI models in place.	Certain revenue generating KPIs in place, ROIs clearly understood	Significant revenue impacts (measured and monitored on a regular basis), initiatives are business case driven	Business strategy and competitive differentiation is based on analytics
Data Governance	Little to None	Initial data warehouse model and architectures	Data definitions and models standardized, enterprise wide metadata management implementation	Clear master data management strategies
Line of Business	Little to None	Visible	Aligned including LOB executives	Cross departmental with CEO level visibility
CIO Engagement	Hidden	Limited	Involved	Transformative

Figure 2-2. *Building up analytical capabilities for big data (Source: www.idc.com)*

Leading practices of enterprise information management for big data platforms

As a set of best practices, we have attempted to give a brief outline of leading practices that will need to be in place to successfully leverage existing information management investments along with implementation of big data platforms.

- **Align big data with specific business goals:** The key intent of big data is to find value from high volumes and varieties of data. It is important to prioritize investments for setting up big data platforms and develop business use cases. Do not make the big data initiative an IT-only fun project.

- **Proactively plan for skill acquisition and development:** One of the biggest obstacles for effective big data management is thought to be an anticipated skills shortage. You can't expect to find these skills in abundance; hence, careful planning and execution needs to be done in effective training, talent management, and information governance areas. Standardizing on an information governance approach will allow enterprises to manage costs and best leverage scarce resources.

- **Optimize capability growth with a big data center of excellence:** A big data center of excellence is an excellent way to scale big data management capability across the enterprise. A big data center of excellence will bring in a collaborative culture among several groups within the enterprise, effectively sharing solution knowledge, enabling effective end user and stakeholder engagement and standardizing on business use cases, governance, communications management, and project management. Irrespective of whether big data management is a new or expanding investment, the soft and hard costs can be an investment shared across the enterprise. Another benefit from this approach is that it will help drive the overall information architecture maturity growth in a more structured and systematic way.

- **Align big data with existing enterprise data management capabilities:** Big data technology proof of concept can perhaps be done on its own as separate pilot projects. However, big data management should be seen as an integral extension of your existing business intelligence and data warehousing platform. This will enable enterprise knowledge workers to correlate different types and sources of data, to make associations, and to make meaningful discoveries. But all in all, big data management should be aligned with existing data management capabilities so as to leverage prior investments in infrastructure, platform, BI and DW.

Implications of Big Data to Enterprise IT?

Big data has many aspects that make it different from traditional data. Big data is generated as a result of user transactions, user interactions, and users browsing behavior. Many sources and channels of big data do not involve human interaction at all; for example, sensor data and machine-generated data. Combined together these types of data make big data poly-structured and hence it does not conform to an easily understood data structure such as that found in transactional data and relational models.

If we just take the structured data in the enterprise, the standard practices to manage enterprise data always had centered on the concept of an "Enterprise Data Warehouse" (EDW). BI tools then work on the data contained in the EDW and produce reports and interfaces that summarize data via multi-dimensional analysis functions (e.g., drill-downs, drill-across, aggregations, time-series analysis, etc.) over various dimensions of data. The design and evolution of an EDW requires a well-conceived data management strategy to bring together relevant data sources from various parts of the enterprise. The multi-dimensional data resides in a comprehensive analysis- oriented data model. Reporting strategies are then developed to leverage the data and then a data governance strategy is implemented to manage and maintain the EDW as a valuable enterprise data asset.

While this EDW approach remains a standard practice, there are several factors like cost, scalability, performance, and ability to handle any type of data, which are beginning to show up as serious shortcomings in the traditional solutions. To a certain extent, the cost and scalability concerns can be addressed by effective usages of commodity hardware and storage solutions, but what about the other data types?

> *So, what additional considerations in the IT stack should be put in place to take into account the big data types?*

Let's first discuss a few fundamental concepts. Across the industry there is a growing opinion that it's not just the volume aspects associated with data that make it difficult to manage. Rather, it is the collection of different types of data, when put together, cannot be processed using conventional methods. Then it becomes a big data situation. What are those "conventional methods"? And why did they suddenly become inadequate? To answer that, it is helpful to understand the type of problems we are trying to solve when we take big data into account.

Right from the mainframe era to client server era, a major expectation from enterprise IT was to ensure that transactional systems (e.g., online transaction processing) ran efficiently, quickly, and consistently. These expectations influenced many technologies and architectures to develop applications using proprietary relational databases on proprietary monolithic servers with proprietary and monolithic storage infrastructures.

Table 2-1. *Various Representations of IT Stacks*

	Traditional IT	Web-Scale Applications	Big Data Analytics Initiatives
Scope	Mostly online transactional processing systems	E-commerce, web sites, search engines	Web search, deep analytics applications
Data Characteristics	Relatively small amount of highly structured data of high quality; small number of users	Combination of structured and un-structured data; millions of users	Massive amounts of structured and un-structured data which is to be analyzed to derive insights, spot trends, etc. Accuracy and insight rather than precision is the key.

(continued)

Table 2-1. (*continued*)

	Traditional IT	Web-Scale Applications	Big Data Analytics Initiatives
Solution Stack	Traditional IT enterprise platforms, mostly ACID compliant.	Less proprietary, open source centric, commodity hardware centric (LAMP Stack – Linux, Apache, MySQL, PHP)	Highly open source centric, scalability, performance is the key, consistency can be compromised (SMAQ Stack – Storage, Map-Reduce, Query)
Database	RDBMS platforms	MySQL	NoSQL
Compute	Proprietary	Distributed processing, Linux on large number of commodity servers	Distributed processing, data is node aware, Linux on large number of commodity servers running map-reduce jobs
Storage	Expensive SANs	Scale out commodity NAS	Scale out commodity NAS, Hadoop compatible file systems

The traditional IT stack (let's define it as "database, storage, and computing") that worked quite well for a relatively small amount of highly valuable and highly structured data began to show serious limitations when faced with a number of challenges. For example, the emergence of web-enabled applications (and millions of user bases) needed cost-effective and innovative approaches to enable distribution of computing and processing of data across large numbers of commodity servers. Let's define this as the LAMP stack (Linux, Apache, MySQL, and PHP). Almost every business is now a "digital business," thus the explosive growth in unstructured data (e.g., text, video, audio, medical images, etc.) all around us. This is why a new stack for IT called SMAQ (storage, map-reduce and query) has emerged.

Let's discuss a few examples to fully understand the implications of the SMAQ stack. Imagine, for example, that you are not only trying to store billions of interactions happening in social chatter boxes but also trying to perform sophisticated analytics on those interactions: such as sentiments expressed by people about a particular brand of product, correlations analysis, and taking these sentiments and linking them to your product sales across stores. Conventional databases can neither handle this kind of scale nor do they have the ability to quickly provide answers to these kinds of questions. Relational databases were designed to maintain transaction history (the ACID principles) in a highly consistent manner and thus inherently they have limitations on scalability and performance. The scale in our example necessitates that we follow a distributed data, storage, and processing approach. The conventional storage and computing approaches can't handle this kind of scale and complexity.

The entire data analysis process starting from data collection to transformations to analysis needs to change to accommodate the distributed design approach. To manage the scale of data, the data set itself needs to be highly distributed. Both the data and the computing for big data scale needs to happen on large numbers of heterogeneous and distributed storage devices. The computing process needs to move closer to the data to reduce I/O cycle times and query results in round trips thus ensuring high performance. In a distributed design approach, the input data is first processed on distributed devices and transformed into intermediate data sets. Then these intermediate results are reduced to aggregated data sets, which is the desired end result. *These two phases of transformation and aggregation are called "map" and "educe."*

In chapters 4, 5, and 6 we will discuss these design approaches at length.

Figure 2-3 provides a high-level view of various components that constitute a SMAQ stack. The SMAQ stack came into existence largely through open-source projects and centers around Hadoop-based architectures and NoSQL databases.

Query	✓ Massively parallel processing of queries ✓ Highly efficient for analytics type of workloads
Map-Reduce	✓ Distributes computing across several servers ✓ Mostly runs as a batch oriented process
Storage	✓ Distributed storage, data is node aware ✓ At a native level data is stored as a file thus enabling schema-less characteristics to manage structured and un-structured data efficiently

Figure 2-3. *SMAQ stack*

Map-reduce

The biggest problem at big data scale was to fit large data sets into a single machine. Traditional approaches like row-chaining and partitioning data were not efficient to handle such large data sets. Map-reduce technology was created at Google to solve the problem of web search indexes over billions of documents. Map-reduce takes a query over a data set, divides it into parallel independent sets of tasks, and runs it in parallel over many nodes. This distribution solves the issue of data too large to fit onto a single machine. The "map" phase processes data item by item and transforms them into intermediate data sets. The "reduce phase" then aggregates these intermediate results into a summarized data set, producing the desired end result. In order to achieve a high degree of parallelism, the only constraint the map-reduce jobs need to adhere to is that the map and reduce tasks must not have any interdependency among the result sets.

In chapters 4 and 5 we will discuss several variations of map-reduce technology at length.

Storage

For map-reduce to work, the data needs to be node aware. In other words, the data needs to be available in a distributed fashion to serve each processing node where map and reduce jobs are executed. The data expected by map-reduce is not stored as we normally store relational data (the entire record at one place); instead relevant data is grouped together and stored in chunks, which are then divided among nodes. Each such data set is identified through key-value pairs.

The standard storage mechanism is a distributed file system having the following characteristics:

- **Fault tolerance:** Since data is distributed across nodes, the storage system should be highly fault tolerant.

- **Extreme scalability:** To accommodate big data scale considerations, the storage system should be highly scalable.

- **Write once and read many times:** The workloads for big data are less transaction oriented and more analysis oriented, hence assuming data will remain unchanged after it is written, there should be provision to achieve high data throughput.

- **Locality of computation:** Moving voluminous data around to do computations introduces severe drags on performance. Instead, moving computation (map-reduce) to data results in faster performance. The file system should have features to facilitate this.

HDFS (the distributed file system in Hadoop-based architecture) provides all the above-mentioned functionalities. Unlike a database, HDFS can store and retrieve data but not index it. So, simple random access to data is not possible through the HDFS. HBase is another component in the Hadoop-based architecture leveraging HDFS as a storage system and provides a column-oriented database designed to store massive amounts of data. Because it creates indexes, HBase offers fast, random access to its contents, though with simple queries. For complex operations, HBase acts as both a source and a sink (destination for computed data) for Hadoop map-reduce.

Hive is another component in the Hadoop-based architecture that provides a data warehousing and analysis-like data store. Hive is built on top of Hadoop providing table-based abstraction over HDFS, which makes it easy to load structured data.

NoSQL databases serve as important components within the SMAQ stack wherein they have built-in map-reduce features that allow computation to be parallelized over distributed data nodes. Hadoop-based systems are most often used for batch-oriented data collection purposes, whereas the NoSQL stores are more aligned to provide faster query response to live applications.

In chapters 5 and 6 we will discuss several of these NoSQL data stores and data modeling approaches at length.

Query

Developing map-reduce jobs to execute your queries is not a trivial task and certainly not as simplistic as writing SQL. To overcome this challenge, SMAQ systems incorporate a higher-level query layer to simplify both the specification of the map-reduce operations and the retrieval of the result.

Query layers typically offer features that handle not only the specification of the computation but also the loading and saving of data and the orchestration of the processing on the map-reduce cluster. Search technology is often used to implement the final step in presenting the computed result back to the user.

In subsequent chapters of this book we will address the approach, strategy, architecture, and technology and skills required for enterprises to build up enterprise information management capability to leverage value from big data and analytics platforms.

End Points

Most enterprises typically have data "islands" often having the same kind of data stored in many repositories. For example, there may be multiple customer-facing applications generating and recording customer interactions and transactions. Multiple systems generating log files may need to be consolidated, aggregated, and analyzed as a whole. What this implies is that there may be data silos within the organization with similar kinds of data at scale. Aggregation of data from these silos is needed to ensure data de-duplication while ensuring that all available data is utilized for analysis. One way to ensure data silo aggregation is through setting up of a big data center of excellence and applying data virtualization techniques.

When big data is analyzed in combination with traditional enterprise data, enterprises can develop a more thorough and insightful understanding of their business, which can lead to enhanced productivity, a stronger competitive position and greater innovation: all of which can have a significant impact on the bottom line.

At a high level, the drivers for an enhanced EIM program are as follows:

- To manage and benefit from massive and growing amounts of data

- Handling uncertainty around data format variability and the velocity of data

- Exploit big data in a timely and cost effective fashion

- Become more agile, despite the constraints of legacy systems

- Tap into huge unstructured data sources for data analysis and business intelligence

- Provide high-performance analytics to support data mining and big data initiatives

To make the most of big data, enterprises must evolve their IT infrastructures to handle the rapid rate of delivery of extreme volumes of data with varying data types, which can then be integrated with an organization's other enterprise data to be analyzed.

Thus, alternative technical solutions are required in order for us to better analyze and utilize the large data sets in the big data scenario.

Big data is a relatively new area and there are quite a number of rapid innovations happening in this space, both in business models as well as technology platforms. Before we delve into the depth of technology and architectures, in the next chapter we will discuss the implications of this new technology for businesses; not all big data and technology architectures are suitable for all types of business use cases.

With increasing industry acceptance of SMAQ stack and big data analytics applications, there are also debates around the usefulness of a "data warehouse" and a strong sentiment to move toward a "data lake". The difference between a data lake and a data warehouse is that in a data warehouse, the data is pre-categorized at the point of entry, which can dictate how it's going to be analyzed. This is especially true in online analytical processing, which stores the data in an optimal form to support specific types of analysis. The problem is that, in the world of big data, we don't really know what value the data has when it's initially accepted from the list of sources available to us. We might know some questions we want to answer, but not to the extent that it makes sense to close off the ability to answer questions that materialize later. Therefore, storing data in some "optimal" form for later analysis doesn't make any sense. Instead, we should be storing the data in a massive, easily accessible repository based on the cheap storage that's available today. Then, when there are questions that need answers, that is the time to organize and sift through the chunks of data that will provide those answers.

References

Big Data Analytics: Future Architectures, Skills and Roadmaps for the CIO: www.idc.com

The next frontier of innovation, competition and productivity: Mckinsey.com

The SMAQ stack for Big Data: http://strata.oreilly.com/2010/09/the-smaq-stack-for-big-data.html

The Big Data DMAQ Down: http://blogs.computerworld.com/18840/big_data_smaq_down_storage_mapreduce_and_query

CHAPTER 3

■ ■ ■

Big Data Implications for Industry

Big Data is not only about the data within the corporate firewalls but also about data outside the firewalls too. Hidden inside these vast reservoirs of data are insights that are waiting to be exploited for favorable business outcomes.

The Opportunity

The interesting aspect of big data is that it enables discoveries across a richer and broader data set. Organizations that established big data analytics platforms and enabled their business users and data analysts to effectively leverage it for decision making have realized significant competitive advantages and opened up new business opportunities.

What exactly is this big data analytics platform? For sure, it isn't as simple as putting a system in place. A big data platform involves a set of technologies (most of it is open source and evolving every day), utilizing programming-driven processes and a management discipline (which is a stark contrast to the bundled in-vendor products for data warehousing and BI solutions). These require different skills that combine business knowledge, deep data analysis, statistical skills, and advanced data visualization skills (which is completely different from the skills surrounding SQL, DBA, data integration and reports development).

It is also interesting to observe companies adopting new business models effectively leveraging big data analytics platforms. Google and Amazon have shown the path. It is not only the web scale companies that are exploiting data but also government organizations that are trying to analyze data to find ingenious ways of collecting tax from defaulters. Health-care companies are becoming proactive in tracking and monitoring their customers' health and design "stay well packages" to reduce health-care costs. The hospitality and travel industry is trying to combine various sources of data including social networking data and create personalized vacation packages for their customers. New companies are developing data mashup technologies combining several aspects of customer behaviors and advising them on what products to buy, when to buy them, and where to buy them, in order to find attractive discount pricing.

Big data creates significant opportunities; but there also exists a significant gap between available data and its effective utilization, and many enterprises are scrambling to address this challenge as shown in Figure 3-1 below.

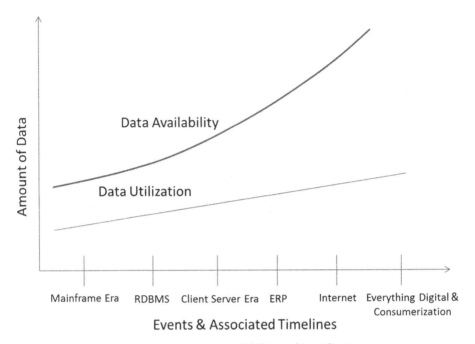

Figure 3-1. *Increasing gap between data availability and its utilization*

Businesses have begun to realize that data is a core asset that not only improves business processes and creates competitive advantage but also enables monetization opportunities. The concern that there is not enough data for analysis is now obviated. The focus has shifted to leveraging data sets to create competitive advantages, and big data analytics platforms are playing a larger role to realize the potential and promise of visionary plans for businesses. With rapid and iterative data insight, business leaders are now making fact-based business decisions. Figure 3-2 plots data sources for big data.

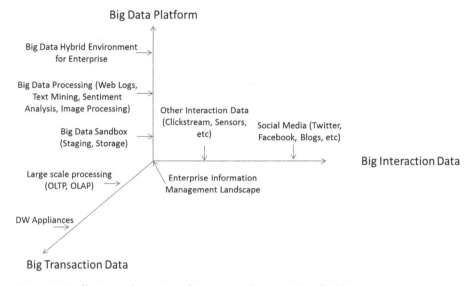

Figure 3-2. *illustrates the various data sources that constitute big data*

Big Data Use Cases by Industry Vertical

Big data and analytics opportunities are not like the traditional data warehousing and BI opportunities where you have a clear road map spanning multiple years. In contrast, the opportunities in the big data and analytics area are business hypothesis driven and often revolve around exploratory activities. The key to exploiting big data analytics is to develop a compelling business use case clearly outlining what e- business outcomes are to be achieved.

The industry-wide use cases shown in Table 3-1, if looked at in isolation across the big data characteristics of volume, velocity, and variety may give you the idea that these problems can be solved using traditional architectures and technology solutions. However, when you see the different aspects of the big data characteristics coming together, you have no choice.

Table 3-1. *illustrates use cases by industry vertical*

Industry	Use Case	Big Data Characteristics		
		Volume	Velocity	Variety
Retail/ ecommerce	• Merchandizing and market basket analysis	✓		✓
	• Campaign management and customer loyalty programs	✓		✓
	• Supply-chain management and analytics	✓	✓	
	• Event- and behavior-based targeting	✓	✓	✓
	• Market and consumer segmentations	✓		✓
	• Recommendation engines— increase average order size by recommending complementary products based on predictive analysis for cross-selling	✓	✓	✓
	• Cross-channel analytics	✓	✓	✓
	• Right offer at the right time	✓	✓	✓
Financial Services	• Real time customer insight	✓	✓	✓
	• Risk analysis and management	✓	✓	✓
	• Fraud detection and security analytics	✓	✓	✓
	• CRM and customer loyalty programs	✓	✓	✓
	• Credit risk, scoring, and analysis	✓	✓	✓
	• High speed Arbitrage trading	✓	✓	✓
	• Trade surveillance, abnormal trading patterns, market manipulation and fraud detection	✓	✓	✓

(continued)

Table 3-1. (*continued*)

Industry	Use Case	Big Data Characteristics		
		Volume	Velocity	Variety
Health & Life Sciences	• Health-insurance fraud detection	✓	✓	✓
	• Campaign and sales program optimization	✓	✓	✓
	• Brand and reputation management	✓		✓
	• Patient care quality and program analysis	✓		✓
	• Drug discovery and development analysis	✓		✓
	• Real-time diagnostic data analysis	✓	✓	✓
	• Research and development	✓		✓
Communication, Media and Technology	• Revenue assurance and dynamic pricing	✓	✓	✓
	• Customer churn prevention	✓	✓	✓
	• Real-time CDR (Call Detail Records) and IPDR (Internet Protocol Detail Records) analysis for network	✓	✓	✓
	• Campaign management and customer loyalty	✓	✓	✓
	• Network performance and optimization	✓	✓	✓
	• Mobile User Location analysis	✓	✓	✓
	• Sentiment analysis, social gaming, online dating, influence, and social graph analysis	✓	✓	✓
Public Sector	• Compliance and regulatory analysis	✓	✓	✓
	• Fraud detection, threat detection, cyber-security, intrusion detection analysis, surveillance and monitoring	✓	✓	✓
	• Smart cities, e-governance	✓	✓	✓
	• Energy consumption and carbon footprint management	✓	✓	✓

(*continued*)

Table 3-1. (*continued*)

Industry	Use Case	Big Data Characteristics		
		Volume	Velocity	Variety
Resources	• Smart grid, smart meters	✓	✓	✓
	• Seismic data analysis	✓	✓	✓
IT Operations	• IT log analysis • Data ingestion of heterogeneous data (structured, semi-structured, poly structured) • Massive write performance • Fast key-value access • Flexible schema and flexible datatypes • Data mash-ups	✓	✓	✓

Data has stories to tell. If data could talk, we might just find out a thing or two about how to run our businesses better. Can you see what your data sees? Can you find out to what it knows? Can you articulate the relationships, patterns, and trends hidden in the massive pile of data?

Some business problems are mathematically compute-intensive, others are more data- analysis intensive, and some are a balance of both. Understanding the nature of the problem is the key to picking the correct approach. The term "big data" is pervasive and has been used to convey all sorts of concepts, including huge quantities of data, social media analytics, next-generation data management capabilities, real-time data, and much more. Whatever the label, organizations are beginning to understand and explore how to process and analyze a vast array of information in new ways.

Many organizations are basing their business cases on the following benefits that can be derived from big data and analytics:

- *Smarter decisions.* Collecting and analyzing new sources of data to not only improve the quality of decision making but also to enable the organization to think beyond the conventional decision-making process.

- *Faster decisions.* Becoming agile and nimble and developing capabilities for the organization to truly become a real-time enterprise. In other words, the decision-making latency is shortened.

- *Impactful decisions.* Delivering business outcomes and business capabilities that are truly unique differentiators.

In the subsequent sections of this chapter, we will touch upon a few use cases by industry and understand what business problems are solved by the big data analytics platform.

Big Data Analytics for Telecom

If there is any industry that has been truly in the thick of unprecedented data growth, it is the telecommunication industry; innovations and offerings like smart phones, mobile broadband services, peer-to-peer information sharing and video-based services have all played significant roles in contributing to this data growth. In addition, the omnipresence of mobility solutions in all aspects of a consumer's life is redefining how products and services are accepted or discarded by customers. Historically, the telecom operators had always had data on their subscribers, their usage patterns, network performance, cell-site information, device level data, as well as billing data and customer-service-related data. However, most of this data resided in the siloed data repositories; they were not organized, and analyzed in a collective way to provide greater insight into customers and their preferences.

The telecom business is also a capital-intensive business, especially in developing and deploying the infrastructure to serve and support an ever-growing consumer base. Thus, a top business priority for a telecom company is to keep delivering new revenue- generating and customer-satisfying services but without overloading infrastructure capacity and network performance and without costs running out of control. In essence, for telecom operators to survive the competition and stay profitable, they need to get a micro-level view of the services they provide, and they need smarter decisions in real time taking into account all critical aspects of their business.

Figure 3-3 illustrates a typical telecom applications landscape overlaid with type of data sources (structured or unstructured) and big data characteristics.

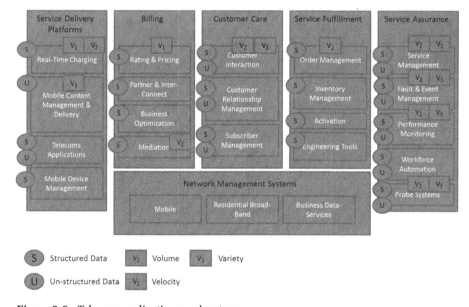

Figure 3-3. Telecom applications and systems

In order to get an enterprise-wide view of their business, the telecom operators had relied on data-integration initiatives moving data from distributed applications to a centralized data repository and then through reporting solutions on top of this data repository. They had developed metrics to review their state of business to identify trends and patterns. But all of this analysis was more or less done in an offline mode, partly due to cost implications of using solutions that can do real-time analysis and secondly due to technology challenges to manage the volume and variety of data. A big data and analytics platform solves these challenges in a cost-effective manner, and below we will discuss a specific use case around improving customer experience.

As a subscriber, the plan you signed up for pretty much defines the services you would get; however, your experience with the services is very dynamic. The primary reason for this is "network performance" and your "usage patterns." Thus, it's all the more important to integrate network performance data with subscriber usage patterns to understand what is happening in the complex intersection of network and services (voice, data, and content). For example, while monitoring the network performance, the telecom operator detects a spike on the load of the network; however, if you don't correlate the network performance issue in real time with the segment of customers who will experience degraded quality of service, you can imagine the kind of a customer service experience you're delivering!

Figure 3-4 below illustrates how a big data analytics platform combines different data types to deliver real-time analytics correlating streaming network data with subscriber data.

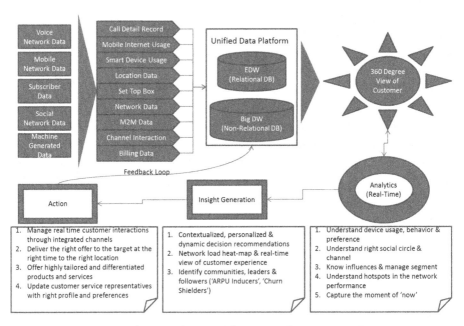

Figure 3-4. *Big data analytics platform to deliver a 360-degree view of the customer at real-time*

The hardest part of collecting and analyzing network data is that it's all semi-structured and streaming in nature. Largely, telecommunication companies have built systems to detect critical outages and bottlenecks and raise alerts, but these systems were rarely designed to analyze network performance over a long period. If you have the ability to understand how and where service issues are trending and how that is affecting your most profitable customers, then you can think of ways to improve.

By combining dropped calls data and latency for video-based services with subscriber's dynamic and static information, you can identify cell towers that are performing poorly and impairing the service experience. This approach can enable operators to analyze and get better insight to network performance and quality of service from a customer's perspective and help them to take proactive measures to answer questions like the following:

- Which regions in my network had the most dropped calls in the past hour, day, week, and which of my customers were most affected? Are these customers profitable? Are they likely to churn?

- Is this a one-off scenario, or it is actually a trend? How can I prioritize where I should invest new capacity in my network, based on customer revenue and profitability?

- Which of the outages were due to handset problems, wireless coverage problems, or switch problems?

- Is my network performance breaching SLAs that have been agreed upon with certain customer segments? How can I prioritize the traffic of those customers in order to avoid SLA breach?

By combining call detail records (CDR) data, cell-site data, calling-circle data, and social network data, you can identify communities and social leaders. This approach can enable operators to quickly determine who are the leaders ("ARPU inducers") and who are the followers ("churn shielders"). These insights can be effectively utilized to develop pricing at an individual level. In addition, the social network monitoring in real time can help in managing "word of mouth" (negative and positive publicity) events based on early detection of social chatters.

Big Data Analytics for Banking

Customer focus is increasingly becoming important for many financial organizations. While customer analytics is not a new concept for banks, some of the best known customer analytics use cases have come from the banking industry: fraud detection, risk analytics, credit scoring, and anti-money laundering are prime examples. However, in the Internet era, the growth of data is posing serious challenges to these customer analytics applications. Banks need new technologies to handle the unprecedented volume, variety, and velocity of information.

Figure 3-5 illustrates a typical banking applications landscape overlaid with the type of data sources (structured or unstructured) and big data characteristics.

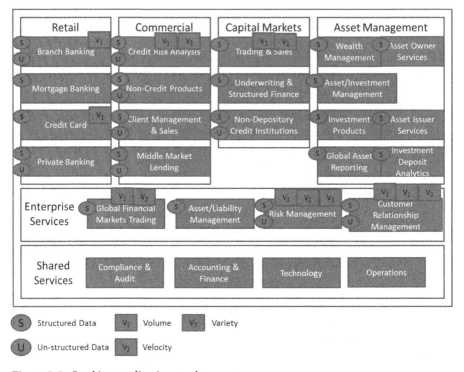

Figure 3-5. *Banking applications and systems*

We will discuss several use cases below.

Next Best Action: "Next best action" is a recommendation engine that takes the bank's business priorities and the customer's needs and comes up with a recommendation to cross-sell, up-sell, or provide a better service to the customer. Next best action is not a new concept for banks; traditionally, the banks have taken their customer transaction data, applied basic segmentation techniques in an offline mode, and devised appropriate offers delivered through marketing channels and customer service representatives. These offers were mostly targeted for customer retention and product promotion campaigns. This approach exposes several drawbacks when you take into account the various channels through which today's customers interact. It is not only the transaction data but also the interaction data that defines the customer's preferences and behavior. Hence, the recommendation engine should take into account all the known information about the customer, including interactions or events, geo-location, channel preference, etc., to arrive at optimal set of recommendations. In addition, the recommendation engine should also consider what the right mode of interaction with the customer is. Based on customer's preferences and historical interaction data, the recommendation engine should advise optimal interaction medium: be it the branch, Web, contact center, ATM, or smart phone.

A big data and analytics platform enables the bank to collect and organize host of additional data such as customer preferences, behavior, interaction history, events and location-specific details, which banks have not previously leveraged, often because the technology to manage volume and variety of data was prohibitively costly.

With a big data analytics platform now you have the capabilities to collect and manage all these data in a cost-efficient manner. Once you have all these data organized, you can then run sophisticated analytics to determine the best set of actions to recommend to a customer, such as a targeted promotion, a cross-sell/up-sell offer, a discounted product or fee, or a retention offer. In addition the big data analytics platform can add value through real-time insight generation and help in faster decision making. The speed of delivering the actionable insights is critical: the faster the bank can take action at the point of interaction, the better the business outcomes.

Predictive Analytics: Banks have been pioneers in predictive analytics, applying statistical modeling techniques on historical data to predict what happens next. Notable examples are: *correlations, back-testing strategies, Monte-Carlo simulations*. However, these predictive models were always run on sampled data. In the current scenario, especially in case of wealth management and capital market functions, these predictive models are fast becoming obsolete and need to be calibrated quite often because the volume and velocity of data are outrunning the usefulness of the predictive models. A big data analytics platform enables the bank with a scalable data ingestion and data storage platform that can keep pace with the volume and velocity of data. Traditionally, predictive analytics was done by running sophisticated algorithms on top of data sets kept in EDWs or analytical data marts. This approach inherently demanded large computing horsepower and also heavy IO contentions. The big data analytics platforms enables new compute and analysis paradigms such as effectively leveraging distributed processing techniques (Map-Reduce) and in-memory computing: in essence, taking the compute workload closer to the data. This approach enables the predictive models to run on the entire data sets, thus providing more insights and at the same time reducing the cycle time to generate insights.

Risk Management: Better risk management is a critical function for banks, everything a bank has to offer (products or services), all revolve around risk. Thus the ability to accurately assess the risk profile of a potential customer or a loan is linked to bank's overall profitability. Credit worthiness assessment is also used to determine specific features to be offered (e.g., credit limit in case of a credit card) at the time of sourcing. Credit worthiness is a dynamic attribute about the customer and it keeps getting updated with new information coming in during the relationship thereafter and remains a very important tool for the credit risk management function of the bank. The ability to rapidly analyze risk scenarios such as aggregation of counter-party exposure across portfolios and customer base fall within the realm of big data analytics. In addition, new sources of data coming from social media are helping in generating new insights about risk profiles of customers. A big data analytics platform not only provides the capability to ingest a variety of data sources but also to deliver data analysis across larger and wider data sets. *Correlating data from multiple and unconnected sources increases the potential to catch fraudulent activities earlier than current methods.* Consider for instance the potential of correlating point-of-sale data (available to a credit card issuer) with web behavior analysis (either on the bank's site or externally), and cross-examining it with other financial institutions or service providers such as First Data or SWIFT. This would not only improve fraud detection but could also decrease the number of false positives.

Retail Banking: Customer centricity is the key to the retail banking business. As retail banking functions are exploring innovative ways to offer new and targeted services to increase customer loyalty, it is increasingly becoming important to look at data sources and analytics capabilities beyond the customer's transactional data. Banks are now collecting and analyzing customer interaction data, location data, and preferences data to develop targeted service offerings with a greater level of sophistication and certainty. Additionally, with the help of a big data analytics platform, banks are now developing complete profiles of their customers, mapping customer life events such as a marriage, childbirth, or a home purchase, which can help banks introduce opportunities for more targeted services and offerings.

Banks have heavily leveraged customer segmentation analytic techniques to devise innovative marketing and sales strategies. However, in the Internet age, the same customer segmentation techniques are turning out to be inadequate. Simplistic segmentations by annual income or funds on deposit are becoming outdated. Two households that exhibit a lot of similarity relative to deposits may actually turn out to be totally different with respect to home equity, credit cards, prepaid debit, etc. Thus, segmentations have to be done at micro level to obtain a much more accurate prediction of needs, attitudes, and buying-spending behaviors. In order to do the fine-grain segmentation, you will need to collect, organize, and correlate a variety of data sources consisting of customer transaction data and customer interaction data. You need to find the right balance between data derived internally and externally. External data can be used to identify customers' financial triggers, pinpointing those who might be new to a geographic area or in the market for a particular financial product. There are data markets that provide balances held at all financial institutions down to the household level and across a variety of categories: deposits, investable assets, investment balances, net assets, mortgage balances, etc. You need leverage of these data sources to develop sophisticated segmentation models to quantify the market share and target the right households to cross-sell.

Let's discuss how the big data analytics platform shown in Figure 3-6 below can help in optimizing operations and improving insights from data. As we move from batch to real-time process integration in financial services, the requirement for real time analytics and an enterprise-wide view of banks operations becomes more acute. Operations managers need to know about the state of operations regardless of client, transaction type, delivery channel, or (in the case of payments) settlement method. In addition to reliable and timely transaction processing, the data produced from the transaction can be more valuable and provide key business insights for operations, marketing, risk management, etc. However, collectively if we look across the various business operations a bank does, the data is very-large volume, often exists in business silos, is structured and unstructured, and exists inside the bank and on the internet (as evident from the various data sources shown in Figure 3-6.

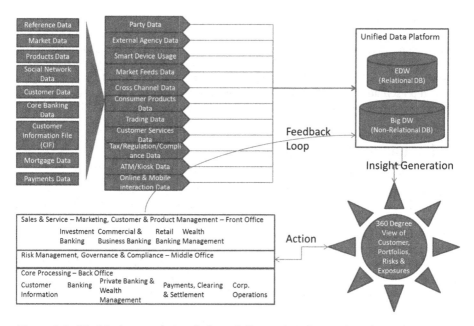

Figure 3-6. *The big data analytics platform delivers a 360-degree view of customer, portfolios, risks, and exposure at real-time*

To demonstrate the value of the big data analytics platform, let's analyze the use case.

Ann, a homeowner is shopping to refinance her home. She begins browsing the Internet to shop for a refinance on her home to take advantage of the favorable interest rates. She does an Internet search and finds options for mortgages and refinancing. Ann selects a bank whose link is near the top of the search results (Her current bank does not show up in the top five of the list at all!). Ann clicks the link to a banking site for more information. When she lands on the banking site, she is presented with an asset of re-finance, which is the topic she searched online for. While she could go directly to this offer, she wants to explore the site and see if she can find more information before applying for the offer. Ann spends a considerable amount of time on the mortgage homepage, and she sees each product contains reviews as well as ratings and comments from other consumers but is unable to make up her mind. Sensing that Ann could be ready to abandon the purchase, an instant chat is offered to help her. Finally, Ann decides on a refinance product and begins an online application process.

Throughout the scenario, you can see different data sources coming into play: clickstream data, customer profile data, browsing behavior data, etc. The big data analytics platform not only puts all of these data in context in one place but offers the ability to correlate, analyze, and provide real-time recommendations. The transactions generated by Ann's refinance scenario are kept in the bank's core systems, the systems that hold these records are operated for high-availability, transactional integrity, data security, and recoverability. The transactions from Ann's branch activity create "events" that initiate actions in near real-time to assess the transaction's validity (for example, anti-fraud measures), and the impact on risk exposure in the lending portfolio. For marketing purposes internal data is combined with external "big-data" from social feeds to monitor social sentiment related to the bank brand. In this case, Ann could be going out there in her social network and talking about her experience with the bank in a very favorable manner. Regardless of channel or different systems in-which data exists, a bank representative can now view a 360-degree profile of Ann, the status of the loan application, the risk profile, her social network profile and comments, etc.

Big Data Analytics for Insurance

Insurance companies have traditionally operated under silos by building their systems and applications serving specific business functions like policy admin, claims, actuary, underwriting, etc. As new processes, products, and technologies emerged, more silos were created due to lack of integration among the existing applications landscape. The traditional methods employed by insurance companies for risk, actuarial and product analyses, reserving, market penetration, customer churn etc., are getting outdated, and business is demanding more modern, accurate, penetrative, and conclusive methods to drive business priorities. Insurers are now actively pursuing analytics in three key areas (customer-centric, risk-centric, and finance-centric), combining internal customer information with new and non-traditional external data sources to provide more granular information of the perceptions and behavior of target audiences.

Figure 3-7 illustrates a typical insurance applications landscape overlaid with types of data sources (structured or unstructured) and big data characteristics.

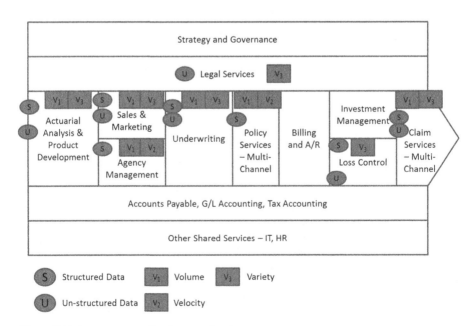

Figure 3-7. *Insurance applications and systems*

The emergence of real-time location data has created an entirely new set of location-based services from navigation to pricing property and casualty insurance based on where, and how, people drive their vehicles. Sensors in the vehicle can now tell how the vehicle is being maintained as well? Insurers are increasingly casting wider nets to collect data beyond the regular insurance data sources (ISO, ACORD, LexisNexis, Marshall-Swift, D&B, Acxiom, comparative raters) to newer data types such as telematics data, social networks, blogs, department of vehicle data, police reports, content streams, geo-spatial and weather patterns. A big data analytics platform provides the capabilities to capture and manage interactions happening through not only the regular distribution channels (agents, brokers, clubs, etc) but also through other multi-channel interfaces like mobile or web based solutions.

Analytics Domains in Insurance

Figure 3-8 shows three analytics domains attributed to the insurance business and associated business opportunities.

Customer Centric	Risk Centric	Finance Centric
• Segmentation • Prospect Identification • Campaign Analysis • Cross Sell/Up Sell • Retention/Lapse • Lifetime Value	• Product Design • Pricing • Underwriting • Telematics • CAT Modeling • Fraud • Reserving	• CAPM • Asset/Liability Matching • Portfolio Optimization • Financial Modeling • Econometric Modeling

Figure 3-8. *Analytics domains and opportunities in insurance*

Risk-centric Analytics: The insurance business is all about understanding risk and becoming better at managing risk. Risk-centric analytics is nothing but assessing the probability of the risk actually happening and expected costs of specific exposures, illnesses, and death. Specialized units within the insurance functions have traditionally developed complex analytical models for product design, pricing, underwriting, loss reserving, and CAT modeling. These analytical models have historically sourced data from the corporate data repositories within the firewalls. However, with the risks and exposures taking new forms, new analytics capabilities are required for insurers to strengthen the core of their business.

To illustrate the point, consider property insurance, where insurers are moving away from coverage-specific risk analysis to more granular by-peril risk analysis. In order to do so, they need to leverage a host of external data on individual perils such as hail, wildfire, coastal storm surge, crime, and dozens of other factors. The time to build and run analytical models to assess all the exposures for individual properties or groups of properties in a portfolio is crucial. The competitive advantage will be lost if it takes weeks or months to go through this process to arrive at a risk based pricing.

For personal auto and commercial auto/fleet, telematics data is becoming a significant source to generate new insights. Data streaming in from telematics devices installed in vehicles is providing a bewildering array of information. Miles driven, location, speed, vehicle performance, driving behavior etc are collected at real time and are used to improve risk assessment, and offer variable risk based pricing strategies.

Customer-centric Analytics: Insurers have always taken the approach of a distributed customer-reach model through agencies, broker networks, self-serving portals, etc. Due to this distributed model of their business, except the core functions like policy admin and claims, all of their customer-centric data resides in multiple places. Therefore, if they need to get a much deeper and more granular understanding of their customer (both at an aggregate levels as well as at individual customer levels), they will have to develop data platforms wherein they can bring in data from disparate sources to create a 360-degree view of their customers. Besides the information available from agents, brokers, and company employees, insurers also need to consider new sources of information such as social media sites and external data on demographics and location-based data. If we take the example of "customer retention" as a business measure, to develop predictive models to identify such cases where a defection or non-renewal is highly probable, you will not only need internal data from CRM, Billing, Policy Admin and Claims systems, you will also need interaction data highlighting relationships and customer behavior patterns.

A big data analytics platform can also increasingly enable insurers to make customer decisions in real time at the time of customer interaction. The big data analytics platform will provide capabilities to analyze web navigation patterns, social media channels of interactions and preferences, and data entry patterns. Understanding these patterns will further help to devise automated or human intervention to prospects/customers.

Finance-centric Analytics: Insurance is a business of risk, hence efficient capital allocation and optimum investment returns are critical to an insurer's financial performance. Insurers frequently use custom-built approaches to develop capital asset pricing models (CAPM) to value and manage assets for least risk and maximum return. Compliance and Regulatory requirements like Solvency II mandates insurers to develop sophisticated models to address areas such as asset/liability matching, investment portfolio optimization, embedded value calculations, and econometric modeling. An increasingly complex business and economic environment is pushing insurers to do more with analytics so that they can dynamically manage the business. Consider the value of being able to combine real-time insight from the operational side of the business with extensive external information concerning macroeconomic attributes and then being able to view risks across portfolios within hours or even minutes.

We end this section with Figure 3-9, showing the analytics architecture and some use cases in the insurance field.

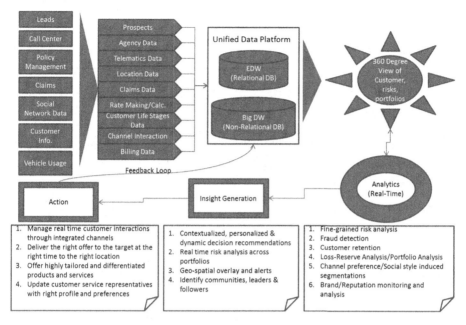

Figure 3-9. Big data analytics conceptual architecture and use cases for insurance

Big Data Analytics for Retail

Historically speaking, retail outlets and supermalls had always generated business by attracting customers to their stores. They had heavily relied on advertisements in various media about attractive pricing, discounts, promotions, etc. They had also invested heavily in setting up physical stores with attractive interior designs, colors, lightning, etc., to provide a wonderful experience to the customers. Customers visiting the brick-and-mortar stores look around for specific products: if they like the pricing, they buy the product. If for some reason they do not like the product or the product is faulty, they call the complaint lines and ask for a refund or exchange.

These outlets and mega-malls built up massive databases by integrating sales data, promotions and campaigns data, supplier data, and in-store inventory data. With this they created customer profiles and started doing high-end analytics like market-basket analytics, seasonal sales analytics, inventory optimization analytics, and pricing optimization analytics. The analytics outcomes provided much valuable insight to business owners regarding their customer behaviors and buying patterns. They developed customer loyalty programs to keep the customer happy.

Figure 3-10 illustrates a typical retail applications landscape overlaid with type of data sources (structured or unstructured) and big data characteristics.

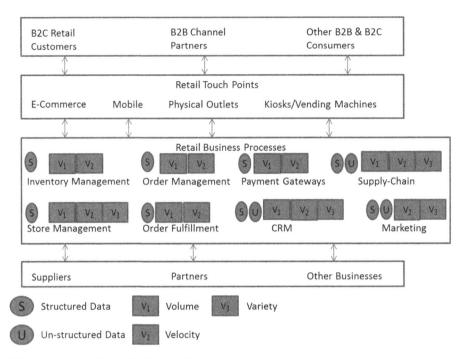

Figure 3-10. *Retail applications and systems*

Every retailer wants to find answers hidden within the massive piles of shopping, spending, inventory, pricing, and clickstream promotion data they have. They want to gain a holistic view of their customers in order to answer business questions such as:

- Who are my customers by categories?

- What are the ways customers buy different product categories with me? How can I use that insight to manage my business in an adaptable manner as my customers switch channels?

- How do my customers behave across a growing number of channels? Should I track customers at the household, individual, digital persona, or touch-point level?

- What are the action-driven behavioral attributes of my customers that best cluster them into segments?

- What is the propensity for my customers to respond across channels and product categories? How can I use that knowledge to interact real-time?

- What is the effectiveness of my current marketing strategy? How can I optimize my investment approach?

- Who is in the market at a given time, what are they looking for, how should I communicate with them, and what is the right positioning strategy for them?

You can very well imagine the number of different types of data sources required to develop a holistic data platform, which can provide answers to the above questions. In addition, retailing is quickly moving from physical presence in store models to multi-channel to multi-screen experiences. First the Internet and then the proliferation of innovative mobile applications are posing significant challenges to the traditional in-store retailing approaches. Simple questions like those above are now becoming extraordinarily complex to answer in a high-velocity shopping environment. The "show rooming" trend is another concern that is increasingly becoming a survival question for the physical-store-based retailers. We will discuss the impact of this trend on retailers, and how a big data analytics platform can help address these concerns.

When customers treat the physical stores as a means to test drive products and then go online with their mobile devices to make transactions at a cheaper price elsewhere, it is called "show-rooming." Increasingly, now mobile and web channels are competing with physical stores as viable alternative channels. A significant percentage of shoppers with smart phones are using applications that can scan QR/barcode codes, get product availability, and compare prices while in the store fully utilizing the concept of *SoLoMoMe* (Social + Local + Mobile + Personalized).

This means retailers need to develop mechanisms to provide unified and integrated customer-centric experiences across all channels of interactions. In other words, retailers need solutions through which they will be able to intelligently interact with customers across traditional and non-traditional channels: websites, physical stores, kiosks, direct mail and catalogs, call centers, social media, mobile devices, gaming consoles, televisions, advertising, home delivery, blogs, and more. This shift in consumer behavior means retailers need to adopt rapid test-and-learn methodologies to deal with the new multi-channel and SoLoMoMe (social + local + mobile data + personalization) reality. But this migration to multi-channel is going to be very difficult for most retailers. The challenge is that retailers have too many physical stores across locations, too little IT integration, and unsustainable cost structures.

Customers generate clues every day as they search, browse, friend, like, tweet, blog, shop, and buy online. But these are massive quantities of data getting generated at a tremendous speed. The challenge for retailers is to develop solutions to capture, analyze, and spot the trends hidden within these massive piles of digital clues before anyone else discovers them. A big data analytics platform is an ideal choice.

Multi-channel analytics is about understanding customer behaviors and presenting new offers via more sophisticated targeting. Figure 3-11 shows the analytics architecture and use cases in the retail space and Figure 3-12 illustrates how market segmentation and customer analytics work together in a big data setting.

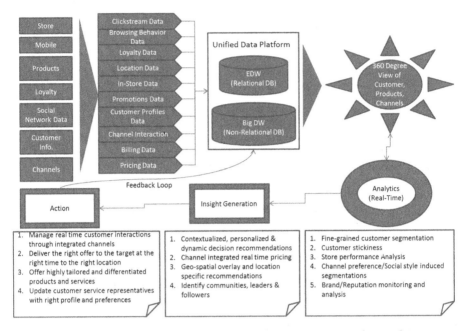

Figure 3-11. *Big data analytics, conceptual architecture, and use cases for retail*

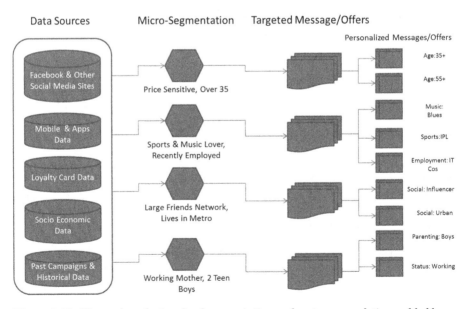

Figure 3-12. *Illustration of micro-level segmentation and customer analytics enabled by big data analytics platform*

You should have the capabilities to ingest data from all possible channels of interaction, analyze the data and through sophisticated analytics and data visualization techniques derive valuable insights. Take for instance, e-mail targeting. The traditional approach has been to scan through your customer base, develop a list of customers to whom you can send appropriate messaging and then send out mass mailers to all. However, today's reality is personalization, by understanding consumers' browsing history you can get down to the point of e-mailing a shopper with a message like "We saw you last night on the women's shoes part of the website" and then send that shopper a targeted shoe promotion. You can also leverage location data from mobile devices; if the customer was in a store but didn't make a transaction then you can mail them a coupon. If the customer opts in, then you can send them a SMS or e-mail promotion code while they are in the store.

By effectively leveraging the big data analytics platform, merchants and marketing teams can gain unprecedented insight into customers' needs and behavior using integrated views around customer shopping and behavior from every touch point and channel. The type of analytics can include: website traffic patterns, traffic by category, traffic by SKU, user demographics, conversion and buying behavior, mobile device patterns (in the case of mobile) and mobile application downloads.

Big Data Analytics for Health Care

Big data has many implications for patients, providers, researchers, payers, and other health-care constituents. The health-care model is undergoing a massive change as the confluence of regionalization, globalization, mobility, and social networking are coming together to voice concerns around increasing cost of health care. From a health-care provider's perspective, the key to profitability was to keep patients in treatment: that is, more inpatient days translating to more revenue. In contrast the new model, which is increasingly supported by government agencies, is to incentivize and compensate health-care providers to keep patients healthy.

In addition, today's patients are demanding more information about their health-care options so that they understand their choices and can participate in decisions about their care. In a health-care scenario, traditionally the key data sources have been patient demographics and medical history, diagnostic, clinical trials data and drug effectiveness index. If these traditional data sources are combined with data provided by patients themselves through social media channels and telematics, it can become a valuable source of information for researchers and others who are constantly in search of options to reduce costs, boost outcomes, and improve treatment.

Governments are increasingly focusing on financial incentives based upon health outcomes. This puts the patient and the patient's care at the center of focus for health-care ecosystem. Both payers and providers are getting incentivized to attain improved outcomes while managing costs. This is a big change for providers who have historically been compensated based upon activity (visits, tests, and treatments). Finally, pharmaceutical companies, which are already paying into health-care reform through a "drug tax," are increasingly getting regulated on drug prices as there is increasing focus on value-based outcomes and with a keen focus on costs.

If all three parties (payer, provider, pharmaceutical company) work collaboratively and share data/insights, disease management programs will become cost-effective and deliver improved patient outcomes at a scale that will further optimize overall health-care cost structure.

Each party brings unique insight, data, and experiences to assist in the design and execution of a health-care management program that could make a sustainable difference.

- Providers bring the deep insight to a patient's health, longitudinal view to the patient's disease progression, and hopefully some historical insight to a patient's past behavior in managing their health.

- Payers bring a comprehensive view to patient medical claims across providers, labs, pharmacies, etc. Additionally, they may have collected one or more health histories to proactively manage at-risk members.

- Although pharmaceutical companies do not bring individual patient data, they do bring a deep understanding of clinical trial data administered on patient populations from both primary and secondary market research studies.

Currently, the health-care solutions do not integrate these different data sources at one place; hence, they lack the ability to do correlations. A big data analytics platform can effectively become the answer. To illustrate the point, let us look at some examples:

- Patients with a chronic condition are identified in the outpatient office for "potential" inclusion in a comprehensive health-care management program.

- The treating health-care professional leverages a set of questions that were developed earlier by patient medical history and geo-demographic data.

- The answers to these questions are then combined with the analytic models to provide a recommendation to the treating health-care professional as to which disease management program level is appropriate for this patient.

- When the patient leaves the outpatient office, depending upon which disease management program he/she was enrolled into, he/she receives follow-up in-home visits, phone calls, letters, email, text messages, and other patient support materials on a continuing basis.

- Additional data from payer, provider, and third party are utilized to initiate follow-ups if the patient does not complete lab tests, attend follow-up appointments, refill their prescriptions, and/or in-home technology that indicates the patient is not being adherent.

Figure 3-13 illustrates a typical health-care applications landscape overlaid with type of data sources (structured or unstructured) and big data characteristics.

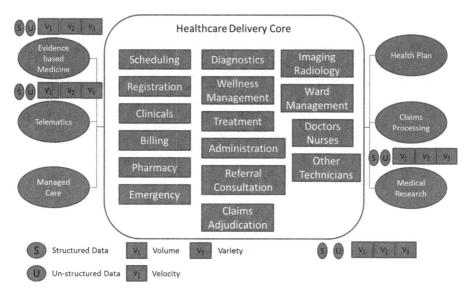

Figure 3-13. *Health-care applications and systems*

This scenario illustrates that by leveraging multiple data sources, the health-care companies can develop efficient and cost-optimized healthcare management programs that are financially sustainable.

There are several other interesting health-care big data use cases that are emerging.

Use case - 1: Keyword mining of doctor's/lab transcripts using text mining and co-relations to patient outcomes.

As a patient interacts with the hospital through multiple diagnosis phases, a lot of information regarding the patient's conditions, diagnosis, and recommendations gets generated. This information is not necessarily of structured data type: for example a CAT scan report would have preliminary interpretation regarding the state of the nerves and blood flow conditions of the brain. There is a host of other types of unstructured text that gets generated as well, such as doctor's remarks, diagnostic lab reports, or nurse's observations. Figure 3-14 illustrated correlations made through text mining with patient outcomes. If these data are effectively leveraged and brought into a Clinical Disease Repository (CDR), we can then apply text analytics to develop early warning signals regarding correlation frequency of occurrence of specific words in the unstructured text and the clinical outcome.

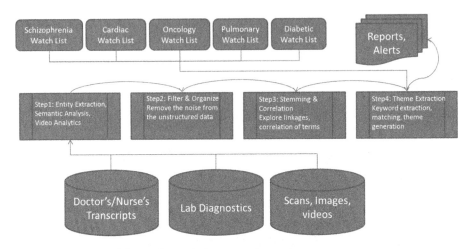

Figure 3-14. *Text mining and correlations to patient outcomes*

Use case - 2: Location aware application analytics for enhancing customer experience and optimizing nurse/doctor deployment.

Hospitals are deploying smart chips to patients, doctors, and nurses to keep track of their whereabouts, the primary reason being so these personnel quickly respond in case of emergencies as shown in Figure 3-15. While at present these smart chips are only issuing location awareness alerts, if this location awareness data can become a new data source it will have huge implications for effectively managing patients experience and optimizing resources within a hospital. For example, we can create models to analyze the strength of the relationships between patient satisfaction index and nurse/patient ratio. We can then define optimal nurse/patient ratios for different sections of the hospital: OPD/cardiology/pediatric wards, for example, may need higher nurse/patient ratios than the dental department. We can define threshold levels to monitor and raise an alert whenever the threshold level is breached to alleviate the risk of an under-serviced patient. We can also use the location awareness data to decide how the various departments must be co-located within the hospital to improve patient outcomes and optimize use of expensive health care equipment.

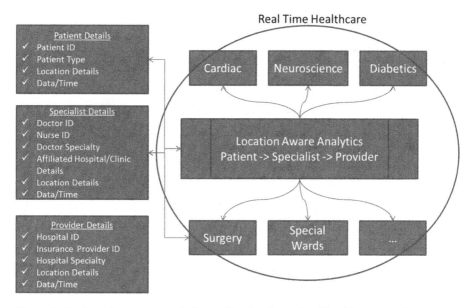

Figure 3-15. *Location aware analytics application for optimal health-care service*

The above use case can also be extended to telemedicine analytics. Telemedicine platforms can be enabled for the patient when it's difficult for the patient to come to the hospital. A telemedicine platform can capture various vitals of the patient like temperature, heart rate, blood pressure and ECG, which can then be streamed into a central repository in real time. Once collated, a series of triggers can be placed on the data to sense and respond to health conditions:

- ALERT-1: If growth in the concentration of BP with statistical significance is found for males in the age group 30 to 45 in a specific zip code, say 08837, then it would be a good idea to hold awareness sessions to sensitize the inhabitants in that zip code to follow healthy eating habits and recommend periodic health checkups.

- ALERT-2: If the number of patient segment migrations > 10 percent based on actual diagnosis events moves from cluster-2 to cluster-5 then proactively import preventive medicine in bulk to cater to growing needs.

Use case - 3: Apriori sequence analysis to define new clinical pathways
A priori algorithms can be used to unearth interesting sequences in data occurring close to each other before a clinical outcome (Figure 3-16). These could be time ordered sequences of events. This would help us create episode rules like *"If 'restlessness' and 'insomnia' occurs in the transcripts there is a 60 percent chance that a coronary episode is imminent."* These can trigger proactive interventions, which can *help* reduce the chances of an adverse event or a hospital admission event.

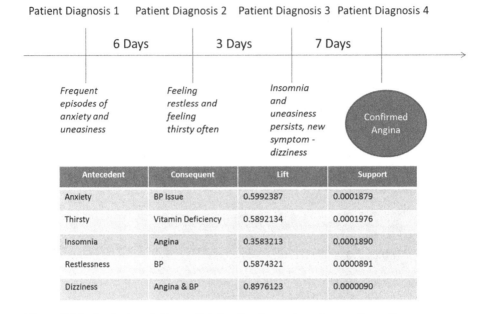

Figure 3-16. *A priori analytics application for diagnosis and preventive actions*

Big Data Analytics for IT/Operations

Since the days of mainframes, IT was always about year-on-year investments on infrastructure, software, skilled people, and process standardizations. IT was primarily responsible to provide a robust platform to support a seven-day per week 9AM to 9PM business model. If the enterprise was running a global business with geographically diverse operations, then the scale of IT operations increased multifold.

First the Internet and then online channels changed the way the IT operations were managed. The business model suddenly shifted to 24 hour/seven days a week, covering all possible time zones. The IT professionals started to use laptops, tablet PCs, and smart devices to interact with enterprise systems wherever and whenever they want. These additional access channels also put a lot of strain on IT standard access policies and controls, and suddenly data privacy and security became a hot topic agenda in CIO discussions.

Hardware and software vendors took advantage of these changing business models and started doing rapid product innovations; it was not uncommon to see multiple vendor products installed in the same enterprise across different lines of businesses. And quite predictably year-on-year IT investment budget kept going up and up.

However, technology evolutions and process maturity started challenging the conventional IT approaches: cloud models, software as a service models (SaaS), hosted service offering models, etc., began to optimize the cost and at the same time provided lot of flexibility on vendor products. It was no longer a single vendor lock-in scenario, rather IT programs evolved to align to business optimization.

IT operations intelligence became a significant contributing factor to changing the paradigm: e.g., running through massive system generated logs to understand potential bottlenecks, alerting the system engineers of an impending system crash, automatically fixing performance issues in the systems, score cards outlining every system's health, and most importantly security and threat detection management.

Underlying the IT transformation is big data and analytics. Analyzing massive amounts of system-generated log data is not a trivial pursuit: these data are cryptic in nature and get generated every nanosecond.

Figure 3-17 illustrates a conceptual architecture that takes in log data, which is mostly voluminous and highly unstructured and delivers several analytics services analyzing the logs and delivering real-time insights.

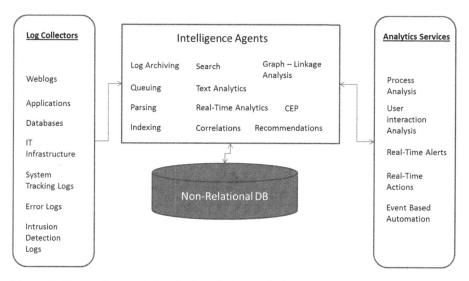

Figure 3-17. *Big data analytics platform for log analysis*

End Points

There's a lot of hype around what a big data analytics platform can deliver, and there is also a host of industry use cases emerging to prove the point. The hardest part of big data analytics is finding an appropriate use case. While every industry is poised to leverage the benefits of big data and analytics, at the same time one needs to understand the fact that the real value from big data and analytics can be derived only if there is a well-thought-out business-relevant scenario. There are many instances of over-enthusiastic technology practitioners chasing elaborate big data technology solutions for a seemingly less significant business outcome use case.

While big data platforms provide a powerful tool to companies as evident from the use cases discussed earlier in this chapter, the process of implementing such a platform and laying down the processes to govern the initiatives is not a trivial pursuit and can very well become a multiyear process. Since the data management approach for big data is different from traditional data management approaches new platforms and methodologies needs to be in place to handle the big data characteristics.

In the next chapter we'll look at the way IT architecture and infrastructure will change examine relevant technology for storage and analysis, and discuss how big data will be analyzed in real time. In addition, we'll also examine the organizational and change management issues that are likely to appear.

References

Big Data Analytics Use Cases: www.practicalanalytics.wordpress.com
Big Data Unlocks Business Values: www.baselinemag.com
http://blog.fluturasolutions.com/2012/12/5-disruptive-big-data-use-cases-in.html
http://blogs.sas.com/content/hls/2011/12/22/disease-management-programs-%E2%80%93-receive-help-from-big-data-and-analytics/
http://www.toolsjournal.com/cloud-articles/item/500-what-are-big-data-use-cases?
Microsoft Industry Reference Architecture for Banking (MIRA-B)May 2012

CHAPTER 4

■ ■ ■

Emerging Database Landscape

Where do newer technologies such as columnar databases and NoSQL come into play? How will you effectively address the impact of big data on application performance, speed and reliability?

In the new data management paradigm and especially considering the influence of big data, IT solutions and enterprise infrastructure landscapes may encompass many technologies working together. Figuring out which of the several technologies are relevant for you is not a trivial matter. In this chapter we will discuss several of these technologies and share best practices: which data management approach is best for what kind of data related challenges?

The ongoing explosion of data today challenges businesses. Organizations capture, track, analyze and store everything from mass quantities of transactional, online, and mobile data, to growing amounts of machine-generated data. In fact, machine-generated data, including sources ranging from web, telecom network and call-detail records, to data from online gaming, social networks, sensors, computer logs, satellites, financial transaction feeds and more, represents the fastest-growing category of big data. High volume websites can generate billions of data entries every month.

Extracting useful intelligence from current data volumes with mostly structured data had been a challenge anyway; imagine the situation when you deal with big data scales!

In order to solve data-volume-related challenges, traditionally architects have applied the below mentioned typical approaches, but each one of the approaches have several implications:

- Tuning or upgrading existing database resulting in significantly increased costs, either through admin costs or licensing fees

- Upgrading hardware processing capabilities increasing overall total cost of ownership to the enterprise (TCO) in terms additional hardware costs and subsequent annual maintenance fees

- Increasing storage capacity, which sets off a recurring pattern: put more storage capacity in direct proportion to the growth of data add incur additional costs

73

- Implementing a data archiving policy wherein old data is periodically moved into lower cost storage solutions. While this is a sensible approach, it also puts constraints on data usage and analysis needs: less data is made available to your analysts and business users for analysis at any one time. This may result in less comprehensive analysis of user patterns and can greatly impact analytic conclusions

- Upgrading network infrastructure leads to both increased costs and, potentially, more complex network configurations.

From the above-mentioned arguments, it is clear that throwing money at your database problem doesn't really solve the issue. Are there any alternative approaches? Before we dive deep into alternative solutions and architectural strategies, let us first understand how databases have evolved over the past decade or so.

The Database Evolution

It is a widely acceptable fact that innovation in database technologies began with the appearance of the relational database management system (RDBMS) and its associated database access mechanism through structured query language (SQL). The RDBMS was primarily designed to handle both online transaction processing (OLTP) workloads and business intelligence (BI) workloads. In addition, a plethora of products and add-on utilities got developed in quick time augmenting the RDBMS capabilities thus developing a rich ecosystem of software products that depended upon its SQL interface and fulfilled many business needs.

Database engineering was primarily built to access data held on spinning disks. The data access operations utilized systems memory to cache data and were largely dependent on the CPU power available. Over time, innovations in efficient usage of memory and faster CPU cycle speeds significantly improved data access and usage patterns. Databases also began to explore options around parallel processing of workloads. During the early days, the typical RDBMS installation was a large symmetric multiprocessing (SMP) server, later these individual servers were clustered with interconnects between two or more servers, thus appearing as a single logical database server. This cluster based architectures significantly improved parallelism, and provided high performance and failover capabilities.

Improvements in the hardware components like memory capacity and network speeds were gradual and continue to evolve. In particular, *in-memory* technology had enabled possibilities of retaining small but frequently accessed datasets in memory. Network speeds also improved to a great extent, making it feasible to assemble much larger clusters of servers known as *grid computing* to further optimize and efficiently distribute workloads. These improvements in the hardware components triggered creation of another type of RDBMS offering known as column-store databases. *Sybase*, now an SAP company, was the first to come out with an enterprise standard database platform, Sybase IQ database.

The column-store databases were designed to address performance issues around query workloads that accessed large volumes of data or large analytical queries, as opposed to row-based databases, which were primarily focused on making sure the transactions, were recorded correctly and quickly in the databases. The biggest push for adoption of column-store databases came from business intelligence applications and analytics applications.

The Scale-Out Architecture

As the data volumes grew exponentially and increasingly there was a need to integrate and leverage a vast array of data sources, a new generation of database products began to emerge. These were labeled as *Not Only SQL* (NoSQL) products. These products were designed to cater to the distributed architecture styles enabling high concurrency and partition tolerance to manage data volumes up to the petabyte range.

Figure 4-1 illustrates scale-out database architecture. You can see the design philosophy where data from several sources are acquired and then distributed across multiple nodes. The full database is spread across multiple computers. In the earlier versions of NoSQL databases there was a constraint that the data for a transaction or query be limited to a single node.

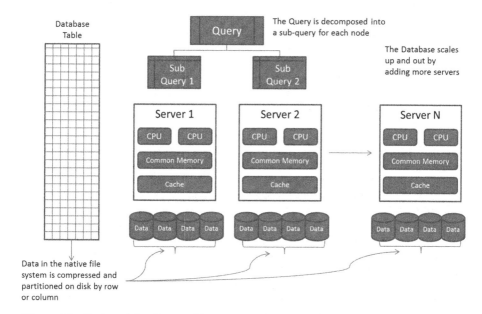

Figure 4-1. *Scale-out database architecture*

The concept of a multi-node database with transactions or queries isolated to individual nodes was a design consideration to support transactional workloads of large websites. Due to this limitation, the back-end database infrastructure of these nodes required manual partitioning of data in identical schemas across nodes. The local database running on each node held a portion of the total data, a technique referred to as *sharding*, for breaking the database up into shards. The queries are broken into sub-queries, which are then applied to specific nodes in a server cluster. The results from each one of these sub-queries are then aggregated to get the final answer. All resources are exploited to run in parallel. To improve performance or cater to larger data volumes, more nodes are added to the cluster as and when needed.

Most NoSQL databases have a scale-out architecture and can be distributed across many server nodes. How they handle data distribution, data compression, and node failure varies from product to product, but the general architecture is similar. They are usually built in a shared-nothing manner so that no node has to know much about what's happening on other nodes.

The scale-out architecture brings to light two interesting features, and both of these features focus on the ability to distribute data over a cluster of servers.

Replication: This is all about taking the same data and copying it over multiple nodes. There are two types of replication strategies.

- Master-Slave

- Peer-To-Peer

In Master-Slave approach, you replicate data across multiple nodes. One node acts as the designated master and the rest are slave nodes keeping copies of the entire data sets, thereby providing resilience to node failures. The master node is the most updated and accurate source for the data sets and is responsible for managing consistency. Periodically, the slaves synchronize their content with the master.

Master-Slave replication is most helpful for scaling when you have a read-intensive data set. You can scale horizontally to handle more read requests by adding more slave nodes and ensuring all read requests are routed to the slaves. However, this approach will have a major bottleneck when you have workloads that are read- and write-intensive, the master will have to juggle around updates and pass on those updates to the slave nodes to make the data consistent everywhere!

While the Master-Slave approach provides read scalability, it severely lacks in write scalability. Peer-to-Peer replication approach addresses this issue by not having a master node altogether. All replication nodes have equal weight, they all accept write requests, and the loss of any of the nodes doesn't prevent access to the data store because rest of the nodes are accessible and have the copies of the same data, although it may not be the most updated data.

In this approach, the concerning fact is about data consistency across all the nodes: when you perform write operations on two different nodes on the same data set, you run into the risk of two different users attempting to update the same record at the same time thus introducing a write-write conflict. This sort of write-write conflicts are managed through a concept called "serialization" wherein, you decide to apply the write operations one after another. Serialization is applied either as pessimistic or optimistic mode. Pessimistic works by preventing conflicts from occurring, in a sense, all write operations are performed in a sequential manner, when all are done, and then only the data set is made available. Optimistic works by letting conflicts occur but detects the instances of conflict and later takes corrective actions to sort them out, making all the write operations eventually consistent.

Sharding: This is all about selectively organizing a particular set of data on different nodes. Once you have data in your data store, different applications and data analysts access different parts of the data set. In such situations, you can introduce horizontal scalability by selectively putting different parts of the data set onto different servers. When the user accesses specific data elements, their queries hit only the designated server. As a result, they get rapid responses!

However, there is one drawback to this approach. If your query consists of data sets distributed over several nodes, how do you aggregate these different data sets? This is a design consideration you need to acknowledge while distributing data over several nodes.

You need to understand the query patterns first and then design the data distribution in such a manner that, data that is commonly accessed together is kept on a single node. This helps in improving query performance.

For example, if you know that most accesses of certain data sets are based on a physical location, you can place that data close to the location where it's being accessed. Or if you see most of the query patterns are around customer's surnames, then you might put all customers with surnames starting from A to E on one node, F to J on another node, like so.

Sharding greatly improves the read and write performance; however, it does little to improve resilience when used alone. Although the data is on different nodes, hence a node failure makes that part of the data unavailable; thus only the users of the data on that shard will have issues, and the rest of the users do not get impacted.

Combining Sharding with Replication: Replication and sharding are two orthogonal techniques for data distribution, which means in your data design considerations; you can use either approach or both the approaches. If you use both the approaches, essentially you are taking the sharding approach but for each shard you are appointing a master node (thus ensuring write consistency); the rest are all slaves with copies of the data items (thus ensuring scalable read operations).

The Relational Database and the Non-Relational Database

On a broad level, we can assume that there are two specific kinds of databases: the relational database and the "non-relational" database. There are several definitions and interpretations of what the characteristics of these two types of databases are.

Let's first define what structured data is and what unstructured data is. These definitions heavily weigh into the characteristics of RDBMS and non-RDBMS systems.

Structured Data: Structured data contains an explicit structure of the data elements. In other words, there exists metadata for every data element and how it will be stored and accessed through SQL-based commands or other programming constructs are clearly defined.

Unstructured Data: Unstructured data constitutes all other data that fall outside the definition of structured data. Its structure is not explicitly declared in a schema. In some cases, as with natural language, the structure may need to be discovered.

The Relational Database (RDBMS): A relational database stores data in tables and pre-dominantly uses SQL-based commands to access the data. Mostly, the data structures and resulting data models take the third-normal form (3NF) structure. In practice, the data model is a set of tables and relationships between them, which are expressed in terms of keys and integrity constraints across related tables such as foreign keys. A row of any table consists of columns of structured data, and the database as a whole contains only structured data. The logical model of the data held in the database is based on tables and relationships.

For example, for an *Employee* table we can define the columns as *Employee_ID, First_Name, Initial, Last_Name, Address_Line_1, Address_Line_2, City, State, Zip_Code, Home_Tel_No, Cell_Tel_No*. In the database schema, we further define the data types for each one of these columns: *integer, char, varchar*, etc. These column names feature in the SQL queries as data of interest for the user. We call this structured data because the data held in the database is represented in a tabular fashion and is known in advance and recorded in a schema.

The Non-Relational Database: Since RDBMS is confined to representing data as related tables made up of rows and columns, it does not easily accommodate data that have nested or hierarchical structures such as a bill of materials or a complex document. Non-relational databases cater to a wider variety of data structures (older mainframe data structures, object and object-relational data structures, document and XML data structures, graph data structures, etc.) than just tables. What we have defined here is an "everything else bucket" that includes all databases that are not purely relational.

OldSQL, NewSQL, and the Emerging NoSQL

The relational database was driven by the idea of database standardization around a generally applicable structure of data to store the data, and a universally acceptable interface like SQL to query the data. We will refer to the traditional RDBMS systems as OldSQL databases. These technologies have proven to be excellent for most transactional data and also for querying and analyzing broad collections of corporate data. These databases are characterized by the use of SQL as the primary means of data access, although they may have other data access features.

There is also a relatively new category of relational databases that although they adhere to the traditional RDBMS philosophy they are designed differently, extending the relational model. A key offering of these databases is new architectures to improve performance, scalability, and most commonly scale-out. They include such products as *Infobright, SAP Sybase IQ, Greenplum, ParAccel, SAND Technologies, Teradata, Vertica, Vectorwise*, and others. We categorize these as NewSQL databases, since they employ SQL as their primary means of access and fundamentally deal with structured data only.

There is also an emerging set of databases specifically designed to provide non-SQL modes of data access. These are commonly categorized as NoSQL databases for their definition of "not only SQL" or "nosql at all." These NoSQL databases exhibit a wide range of characteristics and design philosophies.

Figure 4-2 illustrates the area of applicability of OldSQL, NewSQL, and NoSQL.

Schema Complexity

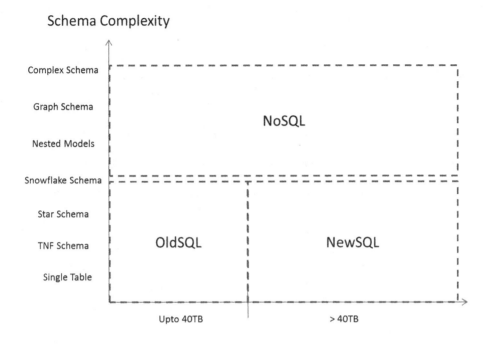

Figure 4-2. *Applicability of OldSQL, NewSQL, and NoSQL*

The vertical axis in Figure 4.2 indicates complexity of data structure. A single table is less complex than the star schema and the snowflake schema structures that one often sees in data warehouses. These are simpler than a third normal form (TNF) relational schema. Nested data, graph data, and other forms of complex data structures represent an increasing complexity of data structures.

It is easy to place OldSQL and NewSQL databases on this diagram. Both cater to all of the data structures up to the snowflake schema models. The distinction between the two categories of product is simply in their ability to scale up to very high volumes of data. The OldSQL databases, built for single server or clustered environments, have a limit to their scalability. Most NewSQL databases, designed for queries over high data volumes, provide little or no support for OLTP, but their scale-out architectures offer good support for data volumes up to the Petabyte level.

As soon as we enter the diverse schema models, NoSQL databases come into the picture. It includes products like *key-value pair databases, graph databases, document databases*, etc. Such databases are built to support extremely large sparse tables and the *JOIN* is superfluous to the intended workloads.

The Influence of Map-Reduce and Hadoop

Until recently there was no widely used framework for parallel programming. Parallel programming was thus a fairly a specialized skill acquired by programmers to develop custom applications. Scale-out hardware models offering parallelism capabilities came onto the scene due to the search engine technology. The Web spans billions of web pages and the number increases daily, yet when searching for a word or phrase you receive an answer in a fraction of a second. This is achieved using parallel computing.

A much publicized use case is Google, each search query that users use, is spread out across thousands of CPUs, each of which has a large amount of memory. A highly compressed schema consisting of the entire web content is held in memory, and the search query accesses that schema. The software framework Google used to address this application is called *Map-Reduce*. Hadoop is another significant component of the parallel computing framework. The Hadoop ecosystem consists of *Hadoop distributed file system* (HDFS), which allows very large data files to be distributed across all the nodes of a very large grid of servers in a way that supports recovery from the failure of any node.

Below is an introduction to map-reduce technology and associated Hadoop ecosystem components. In Chapter 5, there will be more about map-reduce and Hadoop ecosystem components.

The map-reduce mode of operation is to partition the workload across all the servers in a grid and to apply first a mapping step (Map) and then a reduction step (Reduce).

- **Map:** The map step partitions the workload across all the nodes for execution. This step may cycle, as each node can spawn a further division of work and share it with other nodes. In any event, an answer set is arrived at on each node.

- **Reduce:** The reduce step combines the answers from all the nodes. This activity may also be distributed across the grid, if needed, with data being passed as well.

In essence, *Hadoop* implements parallelism that works well on large volumes of data distributed across many servers. The processing is kept local to each node (in the Map step), and only sent across the network for arriving at an answer (in the Reduce step). It is easy to see how you can implement an SQL-like query using this, since the Map step would do the *SELECT* operations, getting the appropriate data from each node, then the Reduce step would compile the answer, possibly implementing a *JOIN* or carrying out a *SORT*.

HDFS keeps three copies of all data by default, and this enables Hadoop to recover from the failure of any of its nodes, as Hadoop also takes snapshots on each node to enable recovery from any node failure. Hadoop is thus "fault tolerant" and the fault tolerance is hardware independent, so it can be deployed on inexpensive commodity hardware. Fault tolerance is important for a system that can be using hundreds of nodes at once, because the probability of any node failing is multiplied by the number of nodes.

In its native form, Hadoop is not a database. *HBase*, another component of the Hadoop ecosystem provides a column-oriented data store capability leveraging Hadoop and HDFS, and it also provides indexing for HDFS. With HBase it is possible to have multiple large tables or even just one large table distributed beneath Hadoop. *Hive* provides a formal query capability turning Hadoop into a data warehouse-like system, allowing data summarization, ad hoc queries, and the analysis of data stored in HBase or at a native level in HDFS. Hive holds metadata describing the contents of files and allows queries in HiveQL, an SQL-like language. It also allows Map-Reduce programmers to get around the limitations of HiveQL by plugging in Map-Reduce routines. *Pig*, originally developed at Yahoo Research, is a high-level language for building Map-Reduce programs for Hadoop, thus simplifying the use of Map-Reduce. It is a data flow language that provides high-level commands.

In summary Hadoop and its ecosystem (*HBase, Hive and Pig*) offer a scale-out database option. Many database companies are using Hadoop in different ways. By leveraging Hadoop's useful capabilities, Aster Data developed a proprietary Map-Reduce environment to extend their relational database, which now complements Teradata's database, after being acquired by Teradata. In their model, a SQL statement can call a Map-Reduce routine to carry out processing and return the results within the context of an SQL statement. This allows existing SQL-compatible tools to make use of Map-Reduce, something that would otherwise require a custom front end.

Other examples include *RainStor*, which uses its compression technology as a Hadoop accelerator; *Cassandra*, which focuses on high volume real-time transaction processing but has integrated Hadoop for batch-oriented analytics; and *MongoDB*, which has a two-way connector that allows for the flow of data between itself and Hadoop. MongoDB could be characterized as a highly scalable document store for web data and is seeing much use by developers for building small custom applications.

Key Value Stores and Distributed Hash Tables

The new generation databases extensively leverage two design philosophies: Key Value Stores and Distributed Hash Tables.

- **Key Value Store:** A key value store is a file that stores records by key, the record consists of a key and attached with it is the data value. The structure of the attached data is not explicitly defined by a schema: in effect it is a blob of data. The primary benefit of such a file is that it is relatively easy to scale in a shared-nothing fashion: it delivers good performance for keyed reads, and developers have more flexibility when storing complex data structures.

- **Distributed Hash Tables:** A distributed hash table (DHT) leverages the key value pair principle but implements a scale-out key value store. Keys are hashed according to their value, so the location of the node on which any given key value pair resides is determined by the hashing process, which distributes the records evenly. The hashing process is itself usually distributed among participating nodes for the sake of failover. Depending on the finer details of the implementation, the outcome is highly scalable since the work of retrieving data can by spread across all participating nodes.

Hadoop's HDFS is a key value store. New generation databases that make use of these techniques include *BerkeleyDB, MongoDB, Riak, Cassandra,* and many others.

XML Defined Data

XML (the eXtensible Mark-up Language) provided an interesting functionality to define metadata along with the data values making the data self-describing. We can refer to such structured data as XML defined data. XML defines data at any level of granularity from a single item through to a complex structure such as a web page.

At this point in time, the use of XML is not as widespread as the use of SQL. However, many of the developer-oriented databases use the *JavaScript object notation* (JSON) rather than XML for data manipulation and interchange.

▪ **Note**   SQL schemas prove to be very useful at the logical level to provide a basis for set-oriented data manipulation but do not define data at the physical level particularly well. The physical definition of data is specific to the database product. XML is broader in some ways as a logical definition of data but is cumbersome at the physical data storage level. JSON, which is object oriented, is less cumbersome than XML at the physical level but lacks logical information about data relationships.

Unstructured Data as Un-modeled Data

If we include XML-defined data in the family of structured data, that still leaves us with a vast amount of data for which no structure has been explicitly declared. This can, in our view, be best designated as "un-modeled data."

This means that it is not possible for the data to be re-used easily by other programs. We classify such data as un-modeled because no design effort has been expended on modeling the data for use by other programs. It may be possible to access the data because we know something about the structure of the data even though it has not been modeled. For example, we may know that some of it is text, so we can search for specific words, and we may even be able to carry out sophisticated searches of this un-modeled data.

It may even be the case that some data held in a typical RDBMS is un-modeled in this way. For example, a specific item in a row may be known (by data type) to be text, but nothing of the inner structure of the text is known. The text itself is not explicitly modeled. Maybe it is just a string of written text or it might be a character stream that defined a web page complete with HTML tags. The database schema does not indicate what it is. Most RDBMSs allow the definition of a *BLOB* (large binary object) and nothing is explicitly defined about the data within the BLOB.

Database Workloads

In the above sections, we discussed how databases have evolved. Given there are so many options and many more to come, it is all the more important to understand which databases are well suited for what kind of workload. This is not a simple criterion because workloads have multiple facets, and there are different architectural approaches to managing these workloads.

Broadly speaking, workload scan be classified into three primary groups:

- **Online Transaction Processing (OLTP):** Transaction processing is a mixed read-write workload that can be become very write-intensive. OLTP requires low latency response, accesses small amounts of data at one time, and has predictable access patterns with few complex joins between different sets of data.

  Streaming data processing and complex event processing type of requirements are at the other extreme end of the OLTP spectrum.

- **Business intelligence (BI):** Originally, this was viewed as a combination of batch and on-demand reporting, later expanded to include ad hoc query, dashboards, and visualization tools. BI workloads are read-intensive, with writes usually done during off-hours or in ways that don't compete with queries. While quick response times are desired, they are not typically in the sub-second range that OLTP requires. Data access patterns tend to be unpredictable; they often involve reading a lot of data at one time, and can have many complex joins.

  Newer concepts like data discovery and exploratory analytics are two other types of workloads where it not only becomes read-intensive but also highly iterative.

- **Analytics:** Analytic workloads involve more extensive calculation over data than BI. They are both compute-intensive and read-intensive. They generally access entire data sets or a combination of different data sets at one time prior to doing computations. Most analytic workloads are done in a batch mode, with the output used downstream via BI or other applications.

Relational databases have been the platform of choice for the above-defined three workloads over the past two decades. As workloads grew larger and more varied, the databases kept adding new features to improve performance. Over the last decade, data volumes and complexity of data types have pushed the workloads past the capabilities of almost all of these RDBMS.

Workload Characteristics

The workload characteristics we are going to discuss below apply more to general database management principles; however, it is essential to understand these characteristics in light of big data and analytics requirements.

What do we mean by workload? Table 4-1 outlines the characteristics of workload in relation to constraints on database.

Table 4-1. Workload Characteristics

Workload Characteristics	Constraints on Database	
	Fewer	More
Read-Write Mix	Low	High
Data Latency	High	Low
Consistency	Eventual	Immediate
Updatability	None	Constant
Data Types	Simple	Complex
Response Times Predictability	High High	Low Low

Different workloads have different characteristics thus posing different challenges when trying to support a mixed workload. Assuming the database is powerful enough to support and is specialized for one particular type of workload is fine, but in the real world there is always a mixed workload.

Read-Write Mix: Whenever we use a database, the workloads are a mix of reads and writes. However, between OLTP, BI and analytics needs you will see these mix of read and writes taking different forms. OLTP is a write-intensive workload whereas BI and analytics are thought of as read-only. Most BI systems write data in bulk at one time and multiple read operations afterward whereas OLTP reads and writes happen at the same time. The intensity of reading and writing and the mix of the two are important aspects of a workload. Business intelligence-specific databases designed to handle read-intensive work are often designed to load data in bulk. While the bulk loads are happening, it is advised not to initiate any other write operations or queries.

Operational BI and dashboards often require up-to-date information. Analytic processing is done in real time as part of the work in OLTP systems. The workload for an operational BI application can look very similar to an OLTP application.

In case of big data scenarios, where many of analytic workloads are based on log data or interaction data, you can expect a high volume of data flowing in continuously, so it must be written continuously. Continuous loading is the extreme end of the spectrum for write intensity. Likewise, in large-scale analytics, particularly when building analytical models, entire data sets are read one or more times, making them among the most read-intensive workloads.

Data Latency: Data latency is the time lag between creation of data and usage of data. Based on the business needs, applications can have different tolerances for latency. For example, OLTP systems have short latencies, with the data available for usage as soon as it has been inserted or updated; whereas data warehouses have long latencies, updated once per day. Short latencies impose more restrictions on a system.

Longer latency requirements mean you have more flexibility in marshaling the database resources. Respite from latency allows you to architect your data management processes in a different way: incremental updates or bulk data processing in a batch mode. The separation of data collection processes from data consumption processes gives flexibility in designing data stores and puts fewer restrictions on a system.

Consistency: Consistency is a critical design consideration. Immediate consistency means that as soon as data has been updated, any other query will see the updated value. Eventual consistency means that changes to data will not be uniformly visible to all queries for some period of time. Some queries may see the earlier value while others see the new or updated value.

Consistency is important to most OLTP systems because inconsistent query results could lead to serious problems. For example, if a bank account is emptied by one withdrawal, it shouldn't be possible to withdraw more funds. If the banking withdrawal application is designed for eventual consistency you can very well imagine the consequences - it might be possible for two simultaneous withdrawals, each taking the full balance out of the account, not a desirable state for the bank.

There are cases where immediate consistency is not critical and eventual consistency is actually a desirable state, as it offers better performance and scalability characteristics, particularly for large scale systems running in a distributed hardware environment like the cloud. For example, in many consumer-facing web applications like e-commerce applications, where the listing of products needs to be consistent with the actual inventory, you can still go ahead with the transaction; later on, products listing can be made consistent with products availability.

Updatability: Data may be changeable or it may be permanent. If an application never updates or deletes data then it is possible to optimize the database design and improve both performance and scalability.

Event streams, such as log data or web tracking activity are examples of data that by its nature does not have updates. Events generate data, systems capture the data and analyze the implications, and the data itself does not undergo any change at all. Outside of event streams, the most common scenarios for write-once data are in BI and analytics workloads, where data is usually loaded once and queried many times thereafter.

A number of BI and analytic databases assume that updates and deletes are rare and use very simple mechanisms to control them. Putting a workload with a constant stream of updates and deletes onto one of these databases will lead to query performance problems because that workload is not part of their primary design. The same applies to some NoSQL data stores that have been designed as append-only data stores to handle extremely high rates of data loading. They can write large volumes of data quickly, but once written the data can't be changed. Instead, it must be copied, modified, and written a second time.

Data Types: Relational databases operate on tables of data, but not all data is tabular. Data structures can be hierarchies, networks, documents, or even nested inside one another. If the data is hierarchical then it must be flattened into different tables before it can be stored in a relational database. This isn't difficult, but it creates a challenge when mapping between the database and a program that needs to retrieve the data.

Response Time: Response time is measured when you execute a query or transaction and the time it takes to return the result of the operation. The challenge with fast response time for queries is the volume of data that must be read, which is itself also a function of the complexity of the query. Many solutions, like OLAP databases, focus on pre-staging data so the query can simply read summarized or pre-calculated results. If a query requires no joins it can be very fast, which is how some NoSQL databases satisfy extremely low latency queries.

Response time for writes is similar, with the added mechanism of eventual consistency. If a database is eventually consistent, it's possible to provide a higher degree of parallelism to partition the workload, offering higher scalability. With proper design, this translates into consistent and low response times.

Predictability: Some workloads have highly predictable data access patterns. For example, OLTP access patterns are usually highly repetitive because there are only a few types of transaction, making them easier to design for and tune for optimal performance. Dashboards and batch reporting will issue the same queries day after day. This repetitive nature allows flexibility in design or tuning of a database since the workload can be anticipated.

When queries are unpredictable, as with ad hoc query or data exploration workloads, the database must be more flexible. The query optimizer must be better so it can provide reasonable performance given unknown queries. Performance management in such scenarios is much more difficult because there is little that can be done in advance to design or tune the workload.

After understanding the different characteristics of workloads, let's examine how they in conjunction with scale (big data characteristics) impose challenges on database technologies. Figure 4-3 shows how the ends of the scale spectrum align with constraints on a database for each characteristic. One or more items on the more restrictive end of the scale can significantly limit the available choice of technologies.

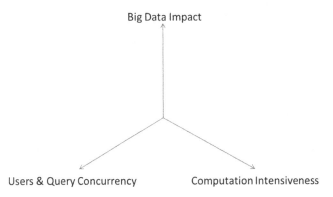

Figure 4-3. *Three axes of big data scale for database processing*

The complexity of a workload and its implication on the database system is defined by a combination of the workload characteristics and the big data scale.

Implication of Big Data Scale on Data Processing

The big data scale is usually represented by big data characteristics, users and concurrency, and computation intensiveness; when combined with varying workloads this can become a handful for evaluation databases.

Big Data Scale: Big data characteristics such as volume, velocity, and variety impart significant implications on database processing (Table 4-2). The simple statistics of database size in gigabytes or terabytes hides many of the important implications. While it is easily understood that increasing volumes of data poses tough challenges to database processing, other characteristics such as streaming data and different data types also add to the complexity of database processing.

Table 4-2. *Big Data Scale: Volume, Velocity, Variety Impact*

Big Data Scale: Volume, Velocity and Variety Impact	
Higher Impact	**Lower Impact**
BI Workloads	OLTP Workloads
Complex Data Structures (Variety)	Simple Data Structures
Many Table Join Operations	Fewer Table Join Operations
High Data Growth Rate	Slow Data Growth Rate
Streaming Data	Mostly Batch Oriented data

Data volume has the biggest impact on BI and analytic workloads as they read large portions of data at one time and join large multiple tables together. In contrast, OLTP workloads are less affected by data volume because the transactions and queries are predictable and involve small amount of data whether recording transaction or fetching records at a time.

There is also a growing demand from businesses to run queries faster, to run more number of queries simultaneously, and to run queries against larger data sets.

To solve the volume-related issues, data management practitioners extensively use compression techniques and indexing. However, there are design-related challenges, as different attributes will have different values: it is therefore not possible to optimize the compression beyond a certain point. Indexes are means to improve query performance but they also introduce additional overheads. Typically, the indexes take up as much space as the data itself, in effect doubling or more. When you add indexes as well as other constructs such as materialized views, the data store size increases as much as eight times the size of the raw data. On the other hand, if you start removing the indexes you see degrading query performance. In effect, while designing the data store, you need to balance the number of indexes.

The structure and complexity of the data can be as important as the raw data volumes. Narrow and deep structures, like simple tables with a small number of columns but many rows, are easier to manage than many tables of varying attributes and row counts. The number of tables and relationships is as important as the amount of data stored. Large numbers of schema objects imply more complex joins and more difficulty distributing the data so that it can be joined efficiently. Variety of data imposes additional constraints on the data store. If your data processing logic needs a combination of structured data and unstructured data, you will now have to design queries to cater to different type of outputs. The resulting data set needs to merge together to provide the expected output. Unstructured data needs parsing, tagging, and filtering techniques to be applied to the raw data; you have to write programs to do these kinds of jobs. The data store itself need to be prepared to accept structured data and unstructured data. These aspects of big data scale drive query complexity, which can result in poor optimizations and lots of data movement.

If your requirement is to cater to streaming data, you will have to take additional considerations while designing your data store. Streaming data need to be captured in real time, and the volume growth aspects associated with streaming data is also critical.

The rate of data growth is important as well. A large initial volume with small incremental growth is easier to manage than a rapidly and unpredictably growing volume of data. Fast growth implies the need for an easily scalable platform, generally pushing one toward databases that support a scale-out model.

There are few helpful rules of thumb for what size qualifies as small or large. In general, when the total amount of data rises to the five-terabyte range, RDBMS databases running on a single server begin to experience performance challenges. At this scale it takes more expertise to tune and manage a system. It is at this boundary that most organizations begin looking to alternatives like purpose-built appliances and parallel shared-nothing databases.

Users Concurrency and Query Concurrency: Concurrency can be defined as the number of simultaneous queries and transactions happening at the same time. In addition, the number of end users accessing the system is often a benchmark to evaluate as far as concurrency is considered. User concurrency can be measured in two ways: passive and active. Active users are those executing queries or transactions, while passive users are the total number of users connected to the database but sitting idle.

With respect to web applications, user concurrency takes a different meaning altogether. The web applications are designed to accommodate unlimited number of users logged into the system at any point of time. Examples are *Facebook, LinkedIn, Gmail,* and *Yahoo*, etc.

In BI workloads, dashboard and scorecard tools may auto-update periodically, making concurrency much higher. In the past it was also reasonable to assume that one user equated to one report and therefore one query. This assumption is no longer true. A dashboard might issue half a dozen complex queries to populate the information on a single screen.

Concurrency is also driven by systems that need to access data in order to execute models, generate alerts, or otherwise monitor data. There are no firm rules for what constitutes high concurrency. The number varies based on workload, since higher workloads have greater impact (see Table 4-3). A dozen concurrent analytics users can stress a database as much as a few thousand users of an OLTP application.

Table 4-3. Big Data Scale User Concurrency Impact

Big Data Scale: User Concurrency and Query Concurrency	
Higher Impact	**Lower Impact**
More distinct users	Less distinct users
More active users and power users	Less active users
Queries spanning across multiple tables of large sizes	Queries spanning across fewer tables of smaller sizes
More scheduled activities	Less scheduled activities

Web applications are less query oriented but more user concurrency oriented. Millions of users log into the system across all time zones. While the transactions they generate may not be volume intensive, the frequency of their transactions is very high: like Tweeting, Facebook status updates, photo uploads/downloads, music sharing, etc. The sheer number of user concurrency and the frequency of transactions impose significant challenges to database processing.

Computation Intensiveness: Computation intensiveness could mean two things: the complexity of the algorithm, or the complexity of the dataset. Running complex algorithms over moderately complex data sets can be a performance challenge. On the other hand, simple algorithms running over large data sets can also cause severe performance issues.

There is no hard and fast definition of what constitutes a complex computation. However, we can say that they typically involve transaction-level data, usually consisting of multiple business rules requiring multiple joins, unpredictable queries, often forced to resort to full table scans. Perhaps a reasonable definition would be that a complex computation always involves multiple set operations. That is, you make a selection and then based on the result of that selection, you go on to make further selections. In other words, complexity involves recursive set operations.

In non-technical terms, complex computations often involve a requirement to analyze and compare different data sets. Some typical complex queries are as follows:

- "To what extent has our new service cannibalized existing products?" – That is, which customers are using the new service instead of the old ones, rather than as an addition.

- "List the top 10 percent of customers most likely to respond to our new marketing campaign."

- "What aspects of a bill are most likely to lead to customer defection?"

- "Are employees more likely to be sick when they are overdue for a holiday?"

- "Which promotions shorten sales cycles the most?"

Consider just the question about the top 10 percent of customers. In order to answer this question we need to analyze previous marketing campaigns, understand which customers responded (which is not easy in itself: it often means a time-lapsed comparison between the campaign and subsequent purchases), and identify common characteristics shared by those customers. We then need to search for recipients of the campaign that share those characteristics and rank them (to do this correctly may require significant input) according to the closeness of their match to the identified characteristics.

It is unlikely that anyone would question the premise that this is a complex computation. You could answer it using a conventional relational database, but it would be time consuming and slow. Now, extend the use case scenario to include all the multi-channel users, and the web scale itself will throw millions of customers into the mix.

Another aspect of complexity is the *predicate*. These are selection criteria such as those based on business rules applied to certain key attributes in a data set. The predicates put considerable pressure on performance and the efficiency of the database.

Unpredictable queries also constitute a large part of the computation issues. Unpredictable queries (as used in exploratory analysis) are, by definition, those where you do not know in advance what the user may want to find out. These impose a number of problems, including:

- You do not know whether the answer to a query will require aggregated data or access to transaction-level data. If the answer can be satisfied by aggregated data then multi-dimensional cubes, materialized views, and pre-defined time-series type of data preparation may be appropriate.

- Even where a query may be satisfied through use of pre-aggregated data, the nature of unpredictable queries is such that you cannot always guarantee that the correct aggregations are available. If they are not then the materialized views and cubes will need to be regenerated or in many cases if the queries are looking for attributes that are not in the materialized views and cubes, then you will have to revert to creation of new pre-aggregation mechanisms.

- The other problem with unpredictable queries is that appropriate indexes may not be defined. While the database optimizer can re-write badly constructed SQL, and determine the most efficient joins and optimize the query path in general, it cannot do anything for lack of indexes. In practice, if a column is not indexed at all, then this will usually mean that the query has to perform a full table scan, and if this is a large table then there will be a substantial performance hit as a result.

■ **Note** One possible option is to build indexes on every conceivable column. Unfortunately this is not usually practical. Every index you build will help to improve the performance of queries that use that index, but at the same time every index you add to the database increases the size of the database.

Certain types of queries require that the whole of a table must be scanned. Some of these arise when there are no available indexes, or from the sorts of complex queries described above. However, very much simpler queries can also give rise to full table scans. For example:

- *"List the full name and email address for customers born in July:"* Given that one in 12 customers are born in July, a typical database optimizer will not consider it worthwhile to use an index, and it will conduct a full table scan. If you have 10 million customers for each of whom you store 3,200 bytes, for instance, then this will mean reading a total of 32,000,000,000 bytes.

- *"Count the married, employed customers who own their own home:"* If we assume the database as above, then conventional approaches still mean reading 32,000,000,000 bytes. In this case, however, column-based approaches can achieve improvements measured in thousands of times.

Another type of data analysis called time-series query brings in significant complexity to the computations. Often you will see workloads that are designed to study people's behavior and activities over time. For example: "Which customers bought smart phones of a certain brand and model within 7 days of ordering a washing machine of the same brand?" In order to answer this sort of query you need to search the database to find out who bought smart phones and then scan for washing machine buying within the required time period.

You cannot easily answer this sort of query using either conventional relational databases or OLAP cubes. In the case of an OLAP solution, you would have to organize your cube by the shortest time period you are ever going to measure against (days in this case) and then you count cells for seven days. Unfortunately, this will mean very large cubes (30 times typical sizes today, which are most commonly implemented by month), and such queries would therefore be extremely inefficient. Moreover, the question posed is based on transaction-level detail in any case, which will not be contained in a cube.

Taken together, the three (big data scale, concurrency, and computation intensiveness) axes define (refer to Figure 4-2) the scale of a workload. Workload scale may grow at different rates along any or all of the axes. The important point to understand when looking at workload scale is that growth along different axes imposes different requirements on a database. Scalability is not one-dimensional.

Database Technologies for Managing the Workloads

Delivering and maintaining good performance isn't a challenge limited to those with hundreds of terabytes of data or hundreds of thousands of users. Many organizations have faced problems with less than a terabyte of data, which can be considered relatively small these days. If good performance is a challenge with moderate data volumes, then why not simply buy more hardware?

Buying more hardware sometimes does solve the problem. However, adding hardware is often a temporary fix because the challenges are due to workloads that the chosen database was not designed to handle. There are usually two ways to scale the database platform: scale up and scale out. Traditional databases are designed to run on a single server with a single operating system. If the server reaches its limit, then the solution is to buy a larger server with more capacity. This is called "scaling up." "Scaling out" means more servers are added to form either a cluster or a grid, with each node running a local database that supports a portion of the workload.

Most organizations approach the initial growth and capacity needs by scaling up their database environment. The problem with this approach is that larger servers are progressively more expensive, whereas the cost is lower for equivalent capacity with several small servers. Eventually scaling up will reach the maximum size of a single server and no more growth will be possible.

Hardware Architectures and Databases

Scaling up at some point becomes unfeasible. The other option is go to for shared-disk database architecture and adding another server running a separate copy of the database software but sharing the same physical storage. This is a half-step toward scale-out architecture because the computers have their own processors, memory, and disk but share a single set of storage.

The challenge with scalability in a shared-disk database architecture is that most databases have limited support for spreading the work of a single query across computers. A shared-disk model will help when the scale problem is concurrency, because more nodes expand the ability to handle growth across discrete tasks.

A shared-disk model will not help if the database limits a query's resource usage to the capacity of a single node, as most shared-disk databases do. If the need is to speed up a small number of large queries, a common need with analytics, then a single query must be parallelizable across more than one node.

Another challenge with the shared-disk model is the shared storage. When all the nodes in a cluster are accessing data, it is possible for the shared disk to become a bottleneck. This is the same problem as running on a single SMP server. There is limited I/O bandwidth between the cluster and the storage.

The I/O bandwidth limit can be partially resolved by increasing the speed and number of storage connections, but these will reach a maximum. At this limit the storage can't deliver data fast enough to meet all the server requests, slowing down queries.

Shared-disk clusters (illustrated in Figure 4-4) improve response time and user concurrency by providing more servers to run queries against the data. They're more appropriate when the problem is the number of concurrent queries and not the need to move large amounts of data for a single query, hence they are less likely to be a good fit for scaling analytic workloads.

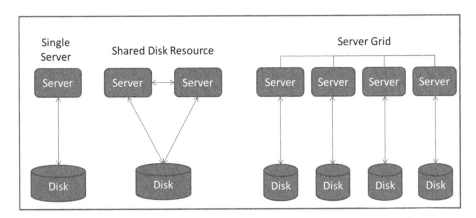

Figure 4-4. Different types of database architectures

An alternate solution taken by many of the newer vendors is to provide a database that can run on a distributed grid of computers with no sharing of components, as shown on the right in Figure 4-4. This architecture is usually called "shared nothing" or *massively parallel processing* (MPP).

In the shared-nothing model each server contains a portion of the database, and no server contains the entire database. It is designed to process as much data possible at each node and share data between nodes only when necessary. Although the database runs independently on multiple nodes, it appears as a single entity to any application.

This model resolves the core limitation of I/O bottlenecks facing single and clustered servers. Adding a node to a shared-nothing database increases the processors and memory available and, more importantly, the disk bandwidth as well. A group of small servers can easily outstrip the total I/O throughput of a very large server or shared disk cluster.

Scaling in this way also lowers the overall hardware cost because commodity servers can be used. A collection of small servers with the same total amount of processors, memory, and storage is less expensive than a single large server. See Table 4-4. We've spent a good deal of time discussing database evolution and various database technologies suitable for different type of workloads. Below are a number of conclusions regarding database architectures:

Table 4-4. *Scale up and scale out considerations*

Scaling up a Database Platform	
Scale Up	**Scale Out**
Vertical expansion/Upgrade to more powerful server configuration	Horizontal expansion through a grid or cluster of commodity servers
More expensive hardware	Less expensive hardware
Eventually hits a limit	Less likely to hit a limit

- RDBMS databases based on the relational model still fit the need for most database implementations, but they have reached scalability limits, making them either impractical or too expensive for specialized workloads. New entrants to the market and alternative approaches are often better suited to specific workloads.

- The relational database is still the preferred choice for most applications today. Database preferences are changing, particularly for new applications that have high scalability requirements for data size or user concurrency. If you find yourself working with a system that has specific needs, let the workload be your primary guide.

- When analyzing the workloads, be sure to consider all the components. For example, if you run a consumer-facing website on the database but also want to analyze data using machine-learning algorithms, you are dealing with two distinct workloads. One requires real-time read-write activity, and the other requires heavy read-intensive and computational activity. These are generally incompatible within the same database without careful design considerations.

Columnar Databases

Organizing data in rows has been the standard approach for so long that practitioners have understood that this is the only way to store and retrieve data. An address list, a customer list, an inventory of products—you can just envision the neat row of fields and data going from left to right on your screen. Databases such as Oracle, MS SQL Server, DB2 and MySQL are the best-known row-based databases.

Row-based databases are ubiquitous because so many of our most important business systems are transactional. Row-oriented databases are well suited for transactional environments, such as a call center where a customer's entire interaction history is required when their profile is retrieved and/or when fields are frequently updated.

Where row-based databases run into trouble is when they are used to handle analytic loads against large volumes of data, especially when user queries are dynamic and ad-hoc in nature.

To understand why, let's look at a database of sales transactions with 50 days of data and 1 million rows per day (Figure 4-5). Each row has 30 columns of data. So, this database has 30 columns and 50 million rows. You want to see how many toasters were sold for the third week of this period. A row-based database would return 7 million rows (1 million for each day of the third week) with 30 columns for each row—or 210 million data elements. That's a lot of data elements to crunch to find out how many toasters were sold that week.

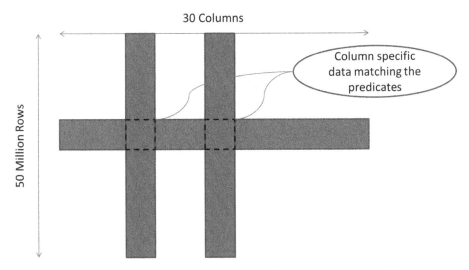

Figure 4-5. Column-based data structure

Column-oriented databases allow data to be stored column by column rather than row by row. Column-oriented databases are better suited for analytics where, unlike transactions, only portions of each record are required. By grouping the data together this way, the database only needs to retrieve columns that are relevant to the query, greatly reducing the overall I/O.

Returning to the example discussed above, we see that a columnar database would not only eliminate 43 days of data, it would also eliminate 28 columns of data. Returning only the columns for toasters and units sold, the columnar database would fetch only 14 million data elements, or 93 percent less data. By returning so much less data, columnar databases are much faster than row-based databases when analyzing large data sets.

Combination/Workload Challenges

The issue here is combining high performance for individual queries with similarly high performance across multiple queries and query types, some of which may be very short running queries and others of which may be long running, or anything in between. Big data scale throws open the environment to address such a combination of workloads. From our discussions around columnar databases, we can see there is a clear architectural benefit to be gained if we use a column-based approach. This is because you don't have to worry about the performance of individual queries and can focus your design efforts to ensure high performance across the potentially (tens of) thousands of queries that may be running at any one time. This is not to say that this is impossible to resolve query performance using the traditional row-based approach, but the challenge is much greater.

Unpredictable Queries: A column is equivalent to an index but without any of the overhead incurred by having to define an index. It's as if you had a conventional database with an index on every column. Suffice it to say, therefore, that if you are undertaking some exploratory analysis using unpredictable queries, then these should run just as quickly as predictable ones when using a column-based approach. Moreover, all sorts of queries (with the exception of row-based look-up queries) will run faster than when using a traditional approach, all other things being equal, precisely because of the reduced I/O overheads.

Complex Queries: Complex queries tend to be slow or, in some cases, simply not achievable: not because of their complexity per se but because they combine elements of unpredictable queries and time-based or quantitative/qualitative queries, and they frequently require whole table scans. Column-based approaches make complex queries feasible precisely because they optimize the capability of the data store in all of these areas.

Large Table Scans: It's usually the case that queries are only interested in a limited subset of the data in each row. However, when using a traditional approach it is necessary to read each row in its entirety. This is wasteful in the extreme. Column-based approaches simply read the relevant data from each column.

Let us take the example of the following query: *List the full name and email address for customers born in July.* Then if the row consists of 3,200 bytes and there are ten million rows then the total read requirement for a conventional relational database is 32,000,000,000 bytes. However, if we assume that the date-of-birth field consists of four bytes, and the full name and email addresses both consist of 25 characters, then the total amount of data that needs to be read from each row is just 54 bytes if you are using a column-based approach. This makes a total read requirement of 540,000,000 bytes.

This represents a reduction of 59.26 times, and this is before we take other factors into account. So it is hardly surprising then that column-based approaches provide dramatically improved performance.

> ■ **Note**   This advantage is not necessarily all one way. Each column you need to retrieve needs to be accessed separately, whereas you can retrieve an entire row in a single read. So the greater the amount of the information that you need from a row the less performance advantage that a column-based approach offers. To take a simplistic example, if you want to read a single row then that is one read. If that row has 15 columns then that is, in theory, 15 reads, so there is a trade-off between the number of rows you want to read versus the number of columns, together with the overhead of finding the rows/columns you need to read in the first place.

A further consideration is that there is a class of query that can be answered directly from an index. These are known as "count queries." Let's take, for example, the question posed previously: *Count the married, employed customers who own a house.* If you have a row-based database, and you have appropriate indexes defined, then you can resolve these queries without having to read the data at all. Of course, in the case of a column-based database the data is the index (or vice versa) so you should always be able to answer count queries in this way.

> ■ **Note**   In a big data environment, count types of queries are common.

Time-based Queries: The issue here is not so much of performance but more of whether relevant queries are possible at all. This is because you not only need the extended SQL in order to handle time-lapse queries but also the ability to store time-stamped transactions. Neither of these is typically the case with traditional RDBMS data stores. Conversely, there are a number of column-based data stores that provide exactly such an approach.

Note that there are a number of use cases that require such capabilities that go beyond conventional databases. For example, in telecommunications it is mandated that companies must retain call detail records, against which relevant queries can be run, often on a time-lapsed basis. Similarly, you will want to be able to run time-based queries against log information (from databases, system logs, web logs and so forth) as well as e-mails and other corporate data that you may need for evidentiary reasons.

Requirements for the Next Generation Data Warehouses

In order to provide the best possible performance to the largest number of users, data warehouses are significantly pre-designed. While logically this may be a reflection of the data model that underpins the data warehouse, in physical terms this means the pre-building indexes, careful partitioning of data, parallel disk striping, developing of pre-aggregated tables, etc.

However, from our discussions so far, we also understood that, the big data scale and type of workloads play a significant role in database design considerations. On the basis of these discussions we can further define some of the features of next generation data warehouses and data platforms:

- Flexibility is paramount: The need is to have a "ask anything" data warehouse.

- It should be possible to store interim results: That is, you may want to perform a query and use the output from that query as a part of the input to another.

- It should be easy to administer, cost effective, and offer a return on investment in as short a timescale as is reasonable.

- It should be efficient in terms of its resources: In particular, the business analyst pursuing a line-of-thought inquiry should be able to pursue any kind of a workload without being constrained by big data characteristics.

- Performance is also fundamental: While different queries will obviously take different lengths of time, typical responses should be in seconds, or minutes at most.

- In modern-day enterprise data warehouses there is a growing requirement to support a much larger number of users/queries than was previously the case and, at the same time, a much broader range of query types.

Data Warehouses and BI systems were built around the notion - data flows from transactional systems possibly through staging areas to ODS to a centralized enterprise data warehouse, the data from the EDW in turn then gets fed into the data marts of various types, which then might feed personal databases. While it was often the case that a single relational database would fulfill many of these data flow needs, this was not always the case, especially where the data of interest is unstructured in nature.

In case of big data, the importance of data design and data flow is all the more critical, as it's evident we'll have to deal with a mix of database technologies and distributed architectures. In addition, we should also do careful considerations around the value of data, as big data by very nature is considered to be full of noises whereas data contained in the EDW is considered to be high quality and important to the organization. Figure 4-6 illustrates a conceptual view of data flow architecture for big data scenarios.

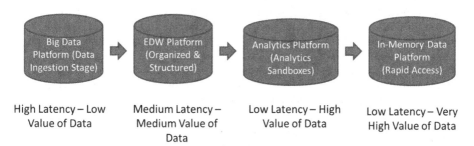

Figure 4-6. *Big data flow*

As the diagram suggests, we might use Hadoop or possibly some NoSQL or NewSQL database to continually gather an ever-growing volume of data, which may be of uncertain data quality. Such data can be characterized as low value, since it is not highly cleansed or processed and may be composed of simple event stream data that requires further processing to derive value.

This data store is analogous to a staging area in traditional data warehouse design but in the big data realm is termed as a "data ingestion" process resulting in a data "lake," whose primary purpose is to support data extracts and transformations intended to feed other data stores. A relatively high latency will usually be adequate for some of this activity. Other uses may require continuous ingest of event streams and real-time monitoring as the data is recorded.

Following this in the data flow is an EDW. Most likely it will serve analytic and BI applications that require a better response time or higher level of concurrency than the data lake could provide. We view this data store as containing more valuable data that has been processed and further enriched and contextualized leveraging data from the data lake.

Following this in the data flow is the analytics sandboxes, wherein you can expect to have a relatively lower level of latency. In this data store, there will be sophisticated analytics modules with very high data computation intensiveness.

Finally, higher value data extracted from the analytic data store flows to an in-memory data store, which feeds applications that demand extremely low latency to satisfy business needs. It may well be the case that the best solution for such a set of business needs is to use different database products for each workload type.

Polyglot Persistence: The Next Generation Database Architecture

Distributed databases and especially NoSQL did solve the scalability and performance- related issues; however, they are just one part of the larger enterprise database management ecosystem. Enterprise database management landscape is all about catering to the mixed workload of OLAP and OLTP. SQL skills and tools are highly prevalent in the enterprise database management ecosystem, and more importantly people have an SQL mind-set. So, assuming a NoSQL-only database management system for the enterprise is a harder fact to accept. The primary challenge with NoSQL is that it's not SQL. Each NoSQL data store is unique and so requires careful design considerations.

SQL focuses on "what" (ability to query the data and use the data) and not "how" (how is the data distributed). Business users and developers are well versed with the "what," now exposing them to also learn the "how" part is increasingly difficult. Hadoop is a great example of this phenomenon. Even though Hadoop has seen widespread adoption it's still limited to silos in organizations. You won't find a large number of applications that are exclusively written for Hadoop. The developers first have to learn how to structure and organize data that makes sense for Hadoop and then write an extensive procedural logic to operate on that data set. The enterprise software is all about SQL. Embracing, extending, and augmenting SQL is a smart thing to do.

But at the same time we can't ignore the power of NoSQL databases, hence the use of heterogeneous data stores within the enterprise is gradually becoming a common practice in application development. Modern applications tend to rely on a polyglot approach to persistence, where traditional RDBMS databases, columnar databases, non-relational data stores, and scalable systems associated with emerging NewSQL and NoSQL technologies, are getting used simultaneously.

So, what exactly is polyglot persistence?

As we have discussed earlier, different databases technologies are designed to solve different workload problems. In addition, there are specialized databases to handle different type of data. Instead of using single database management software for all of the enterprise data requirements, it is wise to look for "horses for courses" approach. This hybrid approach of mixing different database technologies and designing database architectures to meet the specifics of business requirements is called "polyglot persistence" (Figure 4-7).

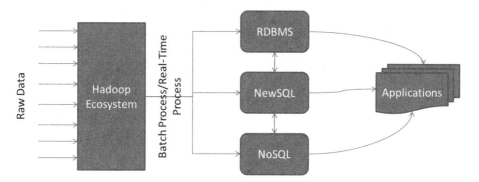

Figure 4-7. *Illustration of a polyglot persistence conceptual architecture*

"Polyglot persistence" refers to the use of both an RDBMS and one or more NoSQL databases as the database management layer for modern applications.
Below, we will discuss few use cases of "polyglot persistence."

How Digg is Built Using Polyglot Persistence

Digg is a social news website. It allows users to discover, share, and recommend web content. Members of the community can submit a webpage for general consideration. Other members can vote that page up ("digg") or down ("bury"). Although voting takes place on digg.com, many websites add "digg" buttons to their pages, allowing users to vote as they browse the Web. The end product is a series of wide-ranging, constantly updated lists of popular and trending content from around the Internet, aggregated by a social network.

The site has several features:

Facebook Connect: Users of Digg and Facebook can connect their accounts. When a Facebook account is connected to a Digg account, Digg articles can then be shared on the user's Facebook page. Facebook Connect also allows Facebook users to log into Digg with their Facebook account, bypassing the need to create a Digg account.

Digg Dialog: Digg users can submit questions to a preselected famous individual who agrees to participate in an interview with a reporter chosen by Digg.

DiggBar: The DiggBar was a frame that gave users access to Digg features without leaving their current webpage. A toolbar above the page allowed users to access Digg comments and analytics.

Digg API: Digg opened their API to the public on April 19, 2007. This allowed software developers to write tools and applications based on queries of Digg's public data, dating back to 2004.

Digg App: Digg released free apps for iPhone and Android in early 2010. The app allowed users to browse stories and digg content. It featured close integration with other social media platforms; users can connect using Facebook or Twitter and share Digg content through them.

To effectively manage all these features, Digg turned to polyglot persistence (see Figure 4-8 below) storing data in multiple database systems depending on the type of data and the access patterns.

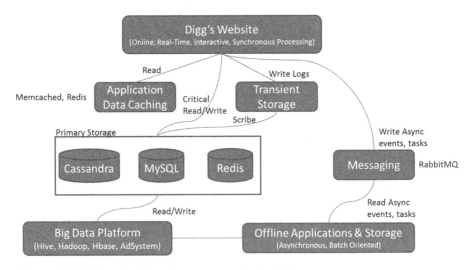

Figure 4-8. *Digg's polyglot persistence conceptual architecture*

- **Cassandra:** Primary storage for access patterns for such things as items (stories), users, Diggs, and the indexes that surround them.

- **HDFS:** Logs user activity from the site and API events. The data source and destination for batch jobs run with Map-Reduce and Hive in Hadoop.

- **MySQL:** Serves as the data store for the story promotion algorithm and calculations, because it requires lots of JOIN heavy operations, which are not a natural fit for the other data stores.

- **Redis:** Primary storage for the personalized news data because it needs to be different for every user and quick to access and update. Redis is used to provide the Digg Streaming API and also for the real-time view and click counts, since it provides super low latency as a memory-based data storage system.

- **Scribe:** This is the log-collecting service. Although this is a primary store, the logs are rotated out of this system regularly and summaries written to HDFS.

Use Case: E-commerce Retail Application

E-commerce stands for "electronic commerce," which is in itself a broad term for selling on the Internet through a website electronically. With the ability to process credit cards electronically on the Internet, just about anything can be sold on the Web. More and more people are enjoying the convenience and lower prices of buying online. Online stores are often able to reduce prices because they are able to eliminate overhead such as employee payrolls required to run a brick-and-mortar store.

E-commerce websites are built differently, but they all use the same basic functions. The customer visits the welcome page and selects a product category. Then the customer browses products within the selected category page and adds a product to his or her shopping cart. The customer continues shopping and selects a different category, then adds several products from this category to the shopping cart. Then the customer selects the "view cart" option and updates quantities for cart products in the cart page. Then the customer verifies the shopping cart's contents and proceeds to checkout. On the checkout page, the customer views the cost of the order and other information, fills in personal data, then submits his or her details. The order is processed and customer is taken to a confirmation page. The confirmation page provides a unique reference number for tracking the customer order, as well as a summary of the order.

What will be our approach, if we apply polyglot persistence to develop the e-commerce application?

It is quite evident that in the e-commerce application in Figure 4-9, we are dealing with different types of data: session data, transaction data, shopping cart data, log level data, product catalogs, customer profile data, etc. It is not necessary for the e-commerce application to use a single data store for all of its needs, since different databases are built for different purposes and not all problems can be elegantly solved by a single database.

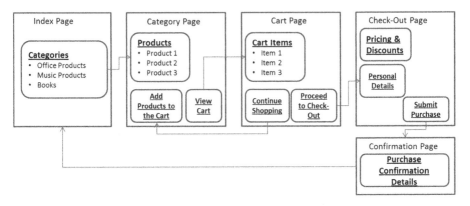

Figure 4-9. *Illustration of process flow for e-commerce application*

Let's discuss each one of these data types and evaluate what kind of a data store will serve the need.

- **Session Data in Redis or Memcached:** Session data requires faster read and write but not durability. For better durability we can always use write-through.

- **Transactional Data in RDBMS:** Order, payment and account in an ACID compliant traditional RDBMS store. In addition to that, HBase/Hive with Hadoop can be used to process transaction-level data such as order history for market basket analysis.

- **Shopping Cart Data in Riak or Cassandra:** A high-availability and fault-tolerance data store such as Riak or Cassandra is the appropriate choice because it is a key/value store with excellent query API with primary key operations such GET, PUT, DELETE, UPDATE.

- **Log Level Data in Cassandra:** Audit and activity in a very high write throughput data store such as Cassandra. This is also good for analytic and real-time data mining such as product ranking, etc.

- **Product Recommendations in Neo4j:** To recommend products to customers when they place products into their shopping carts—for example, "your friends also bought these products" or "your friends bought these accessories for this product"—then introducing a graph data store in the mix becomes relevant. Related products and similar products are in a graph database such as Neo4j.

- **Product Catalogue in MongoDB:** A document-oriented data store that will provide high-read throughput and the ability to handle frequent data change (stock-level information).

- **Customer Profile Data in MongoDB:** A document-oriented data store to manage purchase history, shipping and billing address, etc.

■ **Note** While there are many advantages with the polyglot persistence approach shown in Figure 4-10, it also introduces complexity in managing such an environment. Each data storage mechanism introduces a new interface to be managed. Furthermore data storage is usually a performance bottleneck, so you have to understand a lot about how the technology works to get decent speed. Using the right persistence technology will make this easier, but the challenge won't go away. Many of these NoSQL options involve running on large clusters. This introduces not just a different data model but a whole range of new questions about consistency and availability.

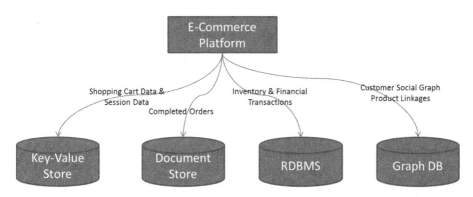

Figure 4-10. *Illustration of a polyglot persistence conceptual architecture*

End Points

Throughout this chapter we have discussed how understanding the workloads is the key to design applications that requires distributed database management functionality. We also discussed that between OldSQL, NewSQL, and NoSQL there are many databases that need evaluation before one embarks on the application development activities. In addition, polyglot persistence is the next new thing. So, what are the best practices and rules that an architect should follow?

The rules that are presented below are in no particular order. A number of these are essentially the same as would apply in a transaction-processing environment, though they may have additional considerations because of the analytic nature of the environment. On the other hand, some of the rules are specific to analytic databases.

Rule 1: Fire and Forget: Application users are only interested in the application they are interfacing with. They do not know about (nor want to know about) the databases that underpin that application. It is thus essential that the characteristics of the databases are invisible to the user, and it should remain that way. This is as true for transactional environments as it is for analytic applications. However, there is an additional consideration when it comes to analytics. In order to get good performance for query-intensive applications you need to focus on performance considerations: index, materialized views, and other such techniques in order to achieve the performance lift.

However, these performance improvement considerations are not feasible to be applied to all possible scenarios. Different workloads demand different kinds of performance improvement considerations; at the time of design, you can't possibly articulate all possible data access scenarios.

For all of these reasons, just choosing a relational database will not be sufficient enough. You will have to look at a combination of databases to solve the business problem.

Rule 2: Ease of Implementation: There are two aspects to this, the first being that the database should be easy to install and, secondly, the resulting application with its underlying database, is easy to deploy. In particular, there should be no requirement for the end user to go through the painful process of understanding the technology architecture components to configure any of the database elements during the implementation process.

Rule 3: High Performance: Everybody expects top-notch performance from their applications; it is all the more important for analytic implementations than for transactional ones. For analytics workloads, it is not easy to predict which particular analysis or complex queries the users are going to run at any particular time. In a transactional environment, on the other hand, you know (roughly) the type of queries that are run and the expected throughput performance that has to be catered to. However, when it comes to analytics there are not only ever-increasing amounts of data available to analyze but also new types of data that might be appropriate to include in queries. Thus, resolving scalability issues is of paramount importance.

Rule 4: High Availability: High availability is always a potential requirement whenever an application is deemed to be "mission critical." The question, of course, is whether analytic applications are regarded as mission critical, and the answer is that it depends on the application and the user. For example, if you have a real-time requirement for security event monitoring, or fraud detection, then high availability is likely to be essential. On the other hand, if you are using analytics to support some sort of customer intelligence application used on a periodic basis, then high availability may not be a critical requirement. The conclusion therefore must be that in some environments it is a must have while in others it is a nice to have. But high availability comes at the cost of other considerations like consistency of data and partition tolerance.

Rule 5: Low Cost: The requirement for low costs is increasingly becoming a hot topic on CIO's agenda. This is not just to the license cost to the software provider but also about the hardware requirements. If you are dealing with big data scales and you need scalability and high availability options, it will be clearly advantageous if the database is highly distributed and runs on low-cost commodity hardware.

Rule 6: Ease of Migration: This won't apply in every case because sometimes new analytic applications are being built rather than existing solutions being ported to a new platform. However, where this does apply, the ease and speed with which the migration can be implemented will be a major factor. There are a number of vendors that support specific capabilities to port from one or other of these environments and ensure the existing applications should run without change and that database schemas can be directly imported into the new environment.

Rule 7: Flexibility: Do you want to offer an environment in which the users can only query what you have pre-prepared for them, or do you want to allow them to make ad hoc or train-of-thought inquiries that go beyond any pre-defined path? While it is always good to provide a comprehensive set of out-of-the-box analytic functions as possible, nobody can predict and design for all possible user data access scenarios. Therefore design for as much flexibility as possible but beware you can't design for all possible scenarios. That's the reason why polyglot persistence is going to become increasingly important.

Rule 8: Loading: There are two circumstances in which the loading capability of the database will be relevant: either because you have large amounts of data to be loaded or because you need to load data in real-time or near real-time. In some circumstances you may have to design for both the scenarios. You will have to evaluate a product that supports a high ingestion rate and (near) real-time capabilities or both. In terms of raw loading capacity this is simply a question of the size of the pipe into the database, bearing in mind any parallelism that is provided. Real-time loading requires support for the ability to micro-batch data (say, batches of one minute) or explicit trickle feeding mechanisms such as change data capture or streaming capability.

Rule 9: Complex Analytics: By "complex analytics" we do not always mean that the questions customers want to ask are complex; but even simple queries (i.e., full table scans, large table joins, etc.) can bring the database to a grinding halt. While there is no formal definition of what constitutes a complex query, they typically involve such things as multi-way and multi-table joins, whole table scans, correlated sub-queries, and other functions that are either computer intensive, I/O intensive or both. Your solution has to be able to perform such queries in a timely manner, and you'll therefore require a database product that can cope with such a workload, also bearing in mind that these queries may be ad hoc and must perform to expectations.

Rule 10: Scalability: If there's one thing you need to be worried about it is increasing volumes of data. Whatever solution you choose needs to be able to easily scale as data volumes grow. It is not just a question of being able to store larger amounts of data; it is also about how quickly you can ingest data from multiple sources. Moreover, it is likely that more queries will be run by more users, as the value of your analytic application or platform becomes apparent to the users, hence the database will also need to be scalable in terms of the user concurrency.

References

THE DATABASE REVOLUTION: A Perspective On Database: Where We Came From and Where We're Going: The Bloor Group

10 Rules: Embedding a Database for High Performance Reporting and Analytics: Bloor Research

martinfowler.com/articles/nosql-intro.pdf

NoSQL Distilled – A Brief Guide to the Emerging World of Polyglot Persistence: Pramod J Sadalage, Martin Fowler

▓ ▓ ▓

CHAPTER 5

Application Architectures for Big Data and Analytics

Big data's bigness is hardly the interesting characteristic. The real fun lies in how we think about data, where they reside in the data ecosystem and how do we generate value from them.

Data-driven platforms are powering business innovations. Generally speaking, transactional data created and stored by enterprise systems such as customer data in CRM applications, operational data in ERP systems, financial data in accounting databases, and sales and marketing data in sales and marketing applications constituted the majority of business relevant data. This data was brought into *enterprise data warehouse* (EDW) systems and BI applications for a consolidated enterprise-wide view and business performance reporting. These system capabilities were gradually challenged due to the scalability and performance considerations purely owing to changing nature of data (volume, velocity, and variety).

Figure 5-1 illustrates the data spectrum along with the big data characteristics. In the earlier chapters we have discussed the need for a radically different approach to data processing and analytics to solve the challenges thrown by big data. In the subsequent sections in this chapter we will discuss what different application architectures we need to consider to implement big data and analytics solutions.

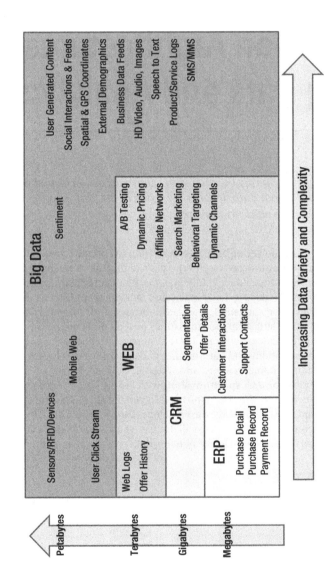

Figure 5-1. Data Landscape

Big Data Warehouse and Analytics

Traditionally, data management and analytics followed a well-governed process (Figure 5-2). First, define business requirements (mostly *metrics/KPIs, reports, data sources*), then develop data integration modules to integrate data from various enterprise systems (CRM, ERP, finance, sales, marketing). Perform data profiling and data quality analysis to certify the correctness and completeness of data, then develop analysis oriented data models (EDW and data marts), then develop reporting and analysis applications (reports, dashboards, multi-dimensional cubes) and finally develop analytics modules (churn models, customer segmentation models, pricing optimization models).

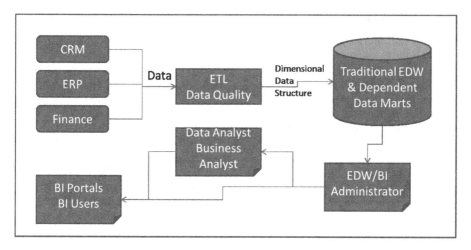

Figure 5-2. *Traditional data processing life cycle*

The influence of the Web, mobile devices, and other technology evolutions challenged this traditional approach and processes primarily due to the changing nature of data. Data is no longer centralized and limited to the enterprise systems only; data has become highly distributed and poly-structured (in many cases loosely structured) and is growing at exponential rates.

Over the last two decades or so, traditional enterprise data warehouses established themselves as the enterprise data asset, but the implementation styles and technologies adopted are very different than the big data scenarios. To leverage big data, most organizations will have to develop an enterprise data platform ecosystem that utilizes traditional EDW data and big data through carefully architected hybrid data warehouse architectures. We call such an ecosystem *big data warehouse* (BDW).

Table 5-1 outlines a comparative view of the business expectations and design principles of big data warehouse (BDW) and enterprise data warehouse (EDW).

Table 5-1. *Comparative View of BDW and EDW Design Principles*

Big Data Warehouse Characteristics	EDW Characteristics
Business Expectations:	Business Expectations:
• Mostly focused on exploratory analysis, finding new insights, quick and ready access to new data, etc. The veracity of results may or may not be questionable since the value of data is not ascertained and not quality controlled upfront.	• Every data entity is strictly governed and quality controlled. Mostly fact based, pre-designed to meet business specific reporting requirements, in many instances the EDW is considered to be single source of truth in the enterprise.
Design Methodology:	Design Methodology:
• Highly agile and iterative approach to enable rapid insights by integrating as many data sources as possible and run as many sophisticated algorithms as possible. Intent is to prove or validate business hypotheses by combining multiple data sources, including internal and external sources, and with or without clear definitions and data models.	• A combination of methodologies (iterative during requirement gathering and reports prototypes creation phases, waterfall during data integration and data model creation phases) are applied to business specific requirements after careful assessment of data needs covering both data integration and data usage aspects. Intent is to provide a consistent, integrated, and single source of truth standardized with data definitions and usage control policies.
Data Architecture Considerations:	Data Architecture Considerations:
• Should have the ability to integrate all possible data structures (both inside the firewalls and external to the corporate). • Should have the ability to scale at relatively low cost. • Should have the ability to analyze massive volumes of data without resorting to sampling mechanisms.	• Not all data is managed and maintained in the EDW, the data sources are previously known, and quality controlled and pre-modeled meeting specific business requirements only. Anything new to be added has to go through a rigorous requirements gathering and validation process. • Has the ability to scale but at a potentially higher cost per byte. • MPP architectures are leveraged to provide performance lift. • Data is periodically archived to accommodate data growth and to keep the cost low.

(continued)

Table 5-1. (*continued*)

Big Data Warehouse Characteristics	EDW Characteristics
Data Integrity and Standards:	Data Integrity and Standards:
• Data integration standards are loosely defined, mostly programmer or application style driven, lack of metadata management, business rules and transformations an integral part of the programs. • Data and data processing programs are highly distributed.	• Driven by relational database management systems principles and architecture approaches (ETL, ELT), data consistency (referential integrities and business rules) and availability drives major development activities. • Data is primarily centralized and data processing programs follow a well-defined execution approach, in most cases these programs are sequential.

Data Design Principles for Big Data Solutions

The distributed nature of big data implies that the data designs must focus on partition-tolerance, secondly to solve the scale issue the data also needs to be distributed across many clusters and nodes hence data designs should also explicitly account for availability. There are two methods broadly applied to address the partition-tolerance and availability requirements:

- Vertical Scaling

- Horizontal Scaling

Vertical Scaling. Vertical scaling simply involves moving the application to larger computers. This approach is also known as "scale up." This works quite well for data but does have limitations such as outgrowing the capacity. It can also be expensive, as you may have to buy newer, bigger, and better machines to cope and this could lead to a vendor lock situation.

Horizontal Scaling. This approach offers more flexibility but is far more complex to manage and design. Horizontal scaling is done by functional scaling, which involves organizing similar data (either through their functional alignment or if some data entities are always queried together) groups and spreading these groups across databases. The second approach is *sharding*, which involves splitting the data within the areas of functionality across multiple databases. This approach is also known as "scale out."

Before we delve deep into data design principles for big data solutions, you should first understand a few established theories governing data design approaches.

ACID. ACID stands for *atomicity, consistency, isolation,* and *durability.* Following *Boyce-Codd's* principles, relational database management systems adopted the ACID approach for data design. In essence, the relational database systems ensured atomicity (a transaction is all or nothing), consistency (only valid data is written to the database), isolation (all transactions are happening serially and the data is correct) and durability (what you write is what you get).

- **Atomicity.** Demands that each transaction is complete: i.e., if one part of the transaction is not successful then the entire transaction will be unsuccessful, and the database will be left unchanged. Atomicity must be guaranteed in each and every circumstance, including crashes, errors, and power failures

- **Consistency.** All parts of a transaction must be consistent in that they must conform to all defined rules, including but not limited to cascades, triggers, constraints, and any combination of these.

- **Isolation.** Ensures that the synchronized execution of transactions results in a system state that can be obtained if transactions are executed serially (i.e., one after another). Each transaction must execute in total isolation (e.g., if transaction 1 and transaction 2 are being simultaneously executed, then each of them should remain unaware of the presence of each other).

- **Durability.** Once committed, a transaction will remain committed even in the event of crashes, errors, or loss. Once the SQL statements within a transaction executed and committed the results to the database, the transaction is stored permanently (even if the database crashes immediately after committing).

ACID relational distributed databases have been a key data design principle for some time; but in today's world of Internet applications (which often needs to be highly scalable due to the huge number of end users and huge data sizes to deal with), there needs to be a different set of design principles, and this is where CAP and BASE theorems came into existence.

CAP. CAP stands for *consistency, availability,* and *partition-tolerance.* CAP theory came into existence because of several shortcomings that came to light when you extend the ACID principles to large distributed data systems.

When you are trying to scale a database system across multiple nodes you run into a few challenges: to make scalable systems that can handle lots of reads and writes you need many more nodes. Now that you are acquiring more and more nodes, reliability becomes a key concern and down time is not acceptable; you need to find a way to handle machine failures. Once you try to scale ACID across many machines you run into performance bottlenecks and network failures. The data processing programs and algorithms run into performance issues in a distributed environment.

CAP theorem addressed these challenges by suggesting that any distributed database system can have, at most, two of the three desirable states: consistency (your data is correct all the time and what you write is what you read), availability (you can read and write your data all the time), partition-tolerance (if one or more nodes fail the system still works and becomes consistent when the system comes online). Typically, CAP is used in relation to consistency in services for high reliability and associated with cache data stores.

- **Consistency.** Nodes at the same time see the same data.

- **Availability.** Guarantee that a response will be given for every request and will show whether the request failed or was successful.

- **Partition Tolerance.** The system will continue to operate despite any failure in a part of the system or random loss of messages.

BASE. This is an alternative to ACID, rather than requiring consistency after every transaction, it is acceptable for the database to eventually be in a consistent state: e.g., it's ok to use stale data, ok to give approximate answers (an example of this is Amazon, does it really matter if the number of books it tells you are available isn't strictly true? The simple answer is no, as even if you do order an item and it then isn't available they refund your money and credit the transaction). BASE is typically associated with NoSQL data stores and it focuses on partition-tolerance and availability and literally puts consistency to a lower priority in order to achieve better partitioning and availability.

In other words, you are designing scalable systems that are *basically available* (system seems to work all the time), *soft state* (it doesn't have to be consistent all the time) and *eventually consistent* (becomes consistent at some later time).

■ **Note** Properties like consistency and availability appear in ACID as well as CAP and BASE theories; however, they differ in implementation approaches because choosing availability as your design consideration for distributed database systems affects some of the ACID principles.

If we reflect back to our earlier discussions around big data characteristics in earlier chapters, it is quite evident that the data will be scattered across multiple nodes and clusters hence the data management system for such scenarios has to be mostly partition-tolerant. Thus, the decision to choose type of database largely depends on what design considerations are important to meet the business use case. If you need a high consistence data model then RDBMS is still the best answer, whereas if you have requirements for high availability and partition-tolerance then NoSQL would be the right choice.

Big Data Warehouse System Requirements and Hybrid Architectures

The system requirements for big data solutions are completely different from traditional data management and analytics solutions. Big data solutions will require a technology or a combination of technologies capable of:

- Managing scale and wide variety of data types covering both the scenarios "data at rest" and "data in motion"

- Managing distributed data across thousands of processors; in many situations the data clusters and grids may be geographically distributed

- Integrating any data source whose structure is not previously known (being schema-read ready)

- Ability to manage and execute workflows that can work across distributed hundreds and thousands of nodes

- Ability to provide built-in semantics to handle and manage trade-offs between consistency, availability and high partition-tolerance functionality

- Ability to support extreme mixed workloads like depth queries as well as breadth queries ranging from ad hoc queries to strategic analysis, and while loading data in batch and streaming fashion

If these are the requirements for big data solutions, do we have any such application architecture that can address all of these requirements? Generally speaking there are two types of application architecture approaches to implement big data solutions: extended RDBMS Architectures extending traditional EDW architectures to manage volume of data and hybrid architectures employing map-reduce/Hadoop architectures to provide a data platform that can manage scale and variety of data types.

A current view of product enhancements of almost all of the major relational database management system vendors outlines an interesting pattern, most of the RDBMS products have significantly evolved adding features like massively parallel processing (MPP) abilities, columnar storage, in-database analytics and ability to execute hadoop map-reduce technologies in the database itself.

This raises a set of interesting questions. How will big data impact your EDW and BI investments? Will it replace them? If not, then how would you combine these two technologies within your current data management architectures?

The intent of these two technologies is different, and their strengths complement each other providing a holistic data platform for enterprises to leverage. The BDW can be used as a data ingestion platform to acquire any type of data of interest at reasonable cost, with little upfront data processing, and less data modeling and data cleansing overheads. The EDW can then utilize these data sources to further enrich the already existing facts and dimensions to support reporting and analytics activities.

The whole point behind the bigness of big data making solutions complex is entirely not true. Consider the scenario *where even 50 GB of data can be said to be big data if the structure is too complex for a normal RDBMS to handle.* In that context, what would we call small data? *Small data* are simple homogenous data structures, e.g. *structured data, strings, dates, times, and all the data we used to feed into the traditional data warehouses.*

Theoretically speaking, a large collection of these small data can eventually become big data.

In any enterprise data management scenario, we will see a combination of small data and big data and there are two application architecture approaches that are widely followed to implement BDW solutions depicted in Figure 5-3.

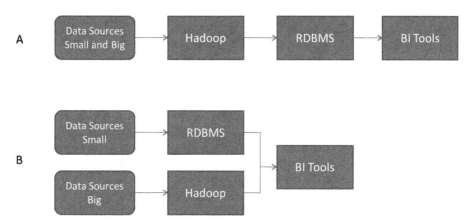

Figure 5-3. *Architecture patterns involving Hadoop and RDBMS*

- The first approach is to have Hadoop as a data ingestion and data processing platform before the data flow reaches the RDBMS.

- The second approach is to have Hadoop as data management platform in parallel to the RDBMS.

In application architecture approach A, Hadoop is used primarily as a data ingestion mechanism and a staging area. In contrast to the normal file system or relational staging area where we can only keep a certain amount of data, using Hadoop as a staging layer we can now keep all the historical data. Apart from historical data, the main advantages of using Hadoop for the staging area are the flexibility to ingest any type of data and also to address scale issues. From the Hadoop staging area we can use specialized data integration tools to move data into RDBMS.

In application architecture approach B, Hadoop is primarily used to store and process data showing big data characteristics, whereas RDBMS is used to store and process "small data." However, both these data stores are used in conjunction to finally make the information available to the consumers.

Enterprise Data Platform Ecosystem – BDW and EDW

A true enterprise data platform should leverage the synergy of the two technology stacks (i.e., BDW and EDW). Together they can provide capabilities to exploit petabyte-scale preprocessing of structured data and unstructured data such as free-form text (e.g., customer comments, user feedback, or product complaints) and semi-structured data such as blogs and click streams. Using map-reduce technologies, the unstructured data and semi-structured data can be transformed to structured results, which can be further fed into the EDW analysis components as new attributes of significance, for example, by first searching for relevant words and concepts and then quantifying the results with counts or other statistics that reveal patterns. These new results can then be combined with other existing facts or dimensions in the EDW to enrich analytic capabilities.

As discussed earlier in this chapter, the goal of BDW is to provide a platform that helps in generating insights by following a discovery type of approach. The EDW data elements (the dimension entities, aggregated facts, enterprise relevant KPIs, metadata information) can significantly enhance this discovery process and shorten the time-to-insight cycle. The conformed dimension tables in EDW reflect a standardized view of business critical entities within the enterprise; they serve as a single source of truth by linking records across several data warehouse fact tables or data marts, they are validated by master data management processes and hence are usually de-duplicated and cleansed. In addition to the dimensional data, other important categories of data within the EDW are hierarchies, metadata, taxonomies, and business rules. These EDW data components may provide a useful business glossary and cross-reference data dictionary during the discovery processes in the BDW. The EDW is also a valuable source of standardized facts, dimensions, and KPIs; these enterprise data elements can be effectively leveraged during the discovery process in the BDW. For example, during the discovery process in BDW you may notice few anomalies in data; but when you reference the metrics in EDW you will realize some of those anomalies are valid business conditions.

Figure 5-4 illustrates the enterprise data platform consisting of hybrid architecture where both the data platforms contribute toward developing an enriched enterprise data store.

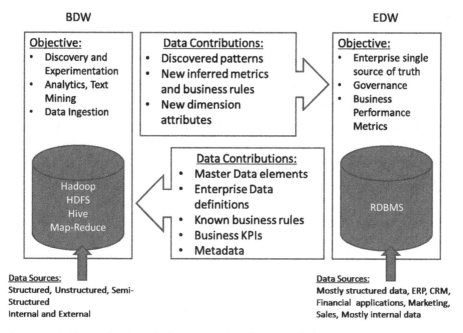

BDW

Objective:
- Discovery and Experimentation
- Analytics, Text Mining
- Data Ingestion

Data Contributions:
- Discovered patterns
- New inferred metrics and business rules
- New dimension attributes

EDW

Objective:
- Enterprise single source of truth
- Governance
- Business Performance Metrics

Hadoop
HDFS
Hive
Map-Reduce

Data Contributions:
- Master Data elements
- Enterprise Data definitions
- Known business rules
- Business KPIs
- Metadata

RDBMS

Data Sources:
Structured, Unstructured, Semi-Structured
Internal and External

Data Sources:
Mostly structured data, ERP, CRM, Financial applications, Marketing, Sales, Mostly internal data

Figure 5-4. Enterprise data platform consisting of BDW and EDW

How does Traditional Data Warehouse processes map to tools in Hadoop Environment?

What we have discussed thus far is a broad view of the enterprise data platform ecosystem consisting of Hadoop and RDBMS components. But these technologies have very distinct characteristics, as described in Table 5-2.

Table 5-2. RDBMS and Hadoop characteristics

Relational DBMSs	Map-Reduce/Hadoop
Mostly proprietary	Open Source
Expensive, Total Cost of Ownership (TCO) grows exponentially	Less expensive, Total Cost of Ownership (TCO) is linear
Data Structures are rigid and needs to be modeled prior	Flexible data structure, less to no modeling required
Great for speedy indexed lookups	Great for massive full data scans
Rich relational semantics	Indirect support for relational semantics, ex: Hive
Indirect support for complex data structures	Deep support for complex data structures
Indirect support for complex algorithms, iterations and branching operations	Deep support for iterations, branching operations and complex algorithms
Deep support for transaction processing	Little to no support for transaction processing

There are several components within the Hadoop environment performing data management operations; below is a listing of their functionality and roles they play, grouped under data management functions as relevant to a data warehouse scenario.

- **Hadoop Distributed File System:** HDFS, the storage layer of Hadoop, is a distributed, scalable, Java-based file system adept at storing large volumes of unstructured data.

- **MapReduce:** MapReduce is a software framework that serves as the compute layer of Hadoop. MapReduce jobs are divided into two parts. The *map* function divides a query into multiple parts and processes data at the node level. The *reduce* function aggregates the results of the map function to determine the answer to the query.

- **Hive:** Hive is a Hadoop-based data warehouse developed by Facebook. It allows users to write queries in SQL, which are then converted to map-reduce. This allows SQL programmers with no map-reduce experience to use the warehouse and makes it easier to integrate with business intelligence and visualization tools such as Micro Strategy, Tableau, Revolutions Analytics, etc.

Hive, initially a sub-project of Hadoop, evolved to provide a formal query capability. In effect, Hive turns Hadoop into something like a data warehouse system, allowing data summarization, ad hoc queries, and the analysis of data stored by Hadoop. Hive holds metadata describing the contents of files and allows queries in HiveQL, an SQL-like language. It also allows map-reduce programmers to get around the limitations of HiveQL by plugging in map-reduce routines.

- **Pig:** Pig Latin is a Hadoop-based language developed by Yahoo. It is relatively easy to learn and is adept at very deep, very long data pipelines (a limitation of SQL.) Pig, originally developed at Yahoo research, is a high-level language for building map-reduce programs for Hadoop, thus simplifying the use of map-reduce. It is a data flow language that provides high-level commands.

- **HBase:** HBase is a non-relational database that allows for low-latency, quick lookups in Hadoop. It adds transactional capabilities to Hadoop, allowing users to conduct updates, inserts, and deletes. E-Bay and Facebook use HBase heavily.

- **Flume:** Flume is a framework for populating Hadoop with data. Agents are populated throughout ones' IT infrastructure (inside web servers, application servers, and mobile devices, for example) to collect data and integrate it into Hadoop.

- **Oozie:** Oozie is a workflow processing system that lets users define a series of jobs written in multiple languages (such as map-reduce, Pig and Hive) then intelligently links them to one another. Oozie allows users to specify, for example, that a particular query is only to be initiated after specified previous jobs on which it relies for data are completed.

- **Whirr:** Whirr is a set of libraries that allows users to easily spin-up Hadoop clusters on top of Amazon EC2, Rackspace, or any virtual infrastructure. It supports all major virtualized infrastructure vendors on the market.

- **Avro:** Avro is a data serialization system that allows for encoding the schema of Hadoop files. It is adept at parsing data and performing removed procedure calls.

- **Mahout:** Mahout is a data-mining library. It takes the most popular data-mining algorithms for performing clustering, regression testing, and statistical modeling and implements them using the map-reduce model.

- **Sqoop:** Sqoop is a connectivity tool for moving data from non-Hadoop data stores such as relational databases and data warehouses into Hadoop. It allows users to specify the target location inside of Hadoop and instruct Sqoop to move data from Oracle, Teradata, or other relational databases to the target.

- **BigTop:** BigTop is an effort to create a more formal process or framework for packaging and interoperability testing of Hadoop's sub-projects and related components with the goal improving the Hadoop platform as a whole.

Clearly, native Hadoop is not a database by any stretch of the imagination. However, once it became popular, it was inevitable that Hadoop would soon evolve to adopt some of the characteristics of a database. HBase, another open source project, stepped in to partially fill the gap. It implements a column-oriented data store modeled on Google's *BigTable* on top of Hadoop and HDFS, and it also provides indexing for HDFS. With HBase it is possible to have multiple large tables or even just one large table distributed beneath Hadoop.

There are a few areas where Hadoop, in its current form, scores well. An obvious one is as an extract, transform, load (ETL) staging system when an organization has a flood of data and only a small proportion can be put to use. The data can be stored in Hadoop and jobs run to extract useful data to put into a database for deeper analysis.

Hadoop was built as a parallel processing environment for large data volumes, not as a database. For that reason, it can be very useful if you need to manipulate data in sophisticated ways. For example, it has been used both to render 3D video and for scientific programming.

It is a massively parallel platform that can be used in many ways. Database capabilities have been added, but even with these it is still best to not think of it as a database product. The open-source nature of Hadoop allowed developers to try it, and this drove early popularity as discussed earlier in Chapter 4. Because it became popular, many vendors began to exploit its capabilities, adding to it or linking it to their databases. Hadoop has generated its own software ecosystem (Figure 5-5).

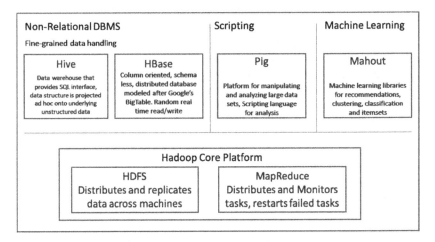

Figure 5-5. *Hadoop conceptual framework*

How Hadoop Works

A client program accesses unstructured and semi-structured data from sources including log files, social media feeds, and internal data stores. It breaks the data up into parts, which are then loaded into a file system made up of multiple nodes running on commodity hardware. The default file store in Hadoop is the Hadoop Distributed File System, or HDFS. File systems such as HDFS are adept at storing large volumes of unstructured and semi-structured data, as they do not require data to be organized into relational rows and columns.

Each part is replicated multiple times and loaded into the file system so that if a node fails, another node has a copy of the data contained on the failed node. A Name Node acts as facilitator, communicating back to the client information such as which nodes are available, where in the cluster certain data resides, and which nodes have failed.

Once the data is loaded into the cluster, it is ready to be analyzed via the map-reduce framework. The client program submits a map job, usually a query written in Java, to one of the nodes in the cluster known as the Job Tracker. The Job Tracker refers to the Name Node to determine which data it needs to access to complete the job and where in the cluster that data is located. Once determined, the Job Tracker submits the query to the relevant nodes.

▓ **Note** The design philosophy is based on the concept that rather than bringing all the data back into a central location for processing, processing occurs at each node simultaneously, or in parallel. This is an essential characteristic of Hadoop.

When each node has finished the processing task, it stores the results. The client program then initiates a reduce job through the Job Tracker in which results of the map phase stored locally on individual nodes are aggregated to determine the answer to the original query, then loaded on to another node in the cluster. The client accesses these results, which can then be loaded into one of number of analytic environments for analysis. The map-reduce job has now been completed (Figure 5-6).

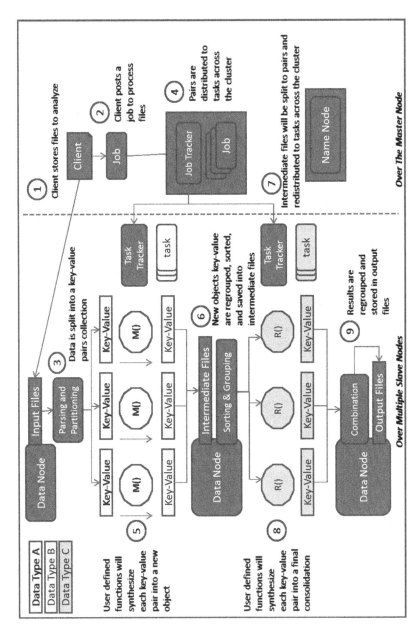

Figure 5-6. Map-reduce job

Eventually an answer is arrived at. The map stage is a filter/workload partition stage. It simply distributes selection criteria across every node. Each node selects data from HDFS files at its node, based on key values. HDFS stores data as a key with attached other data that is undefined (in the sense of being in a schema). Hence it is a primitive key value store, with the records consisting of a head (the key) and a tail (all other data).

The map phase reads data serially from the file and retains only keys that fit the map. Java hooks are provided for any further processing at this stage. The map phase then sends results to other nodes for reduction, so that records that fit the same criteria end up on the same node for reduction. In effect, results are mapped to and sent to an appropriate node for reduction.

The reduce phase processes this data. Usually it will be aggregating or averaging or counting or some combination of such operations. Java hooks are provided for adding sophistication to such processing. Then there is a result of some kind on each reduce node.

Further reduction passes may then be carried out to arrive at a final result. This may involve further data passing in the form of mapping and reducing, making up the full Hadoop job. In essence, this is simply a parallelization by workload partitioning scheme with the added nuance of being fault tolerant.

Once the map-reduce phase is complete, the processed data is ready for further analysis by data scientists and others with advanced data analytics skills. Data scientists can manipulate and analyze the data using any of a number of tools for any number of uses, including to search for hidden insights and patterns or to use as the foundation to build user-facing analytic applications. The data can also be modeled and transferred from Hadoop clusters into existing relational databases, data warehouses, and other traditional IT systems for further analysis and/or to support transactional processing.

Hadoop Technical Components

A Hadoop "stack" is made up of a number of components. They include:

- **Hadoop Distributed File System (HDFS):** The default storage layer in any given Hadoop cluster;

- **Name Node:** The node in a Hadoop cluster that provides the client information on where in the cluster particular data is stored and if any nodes fail.

- **Secondary Node:** A backup to the Name Node, it periodically replicates and stores data from the Name Node should it fail.

- **Job Tracker:** The node in a Hadoop cluster that initiates and coordinates map-reduce jobs, or the processing of the data.

- **Slave Nodes:** The grunts of any Hadoop cluster, slave nodes store data and take direction to process it from the Job Tracker.

In addition to the above, the Hadoop ecosystem is made up of a number of complimentary sub-components. NoSQL data stores like Cassandra and HBase are also used to store the results of map-reduce jobs in Hadoop. In addition to Java, some map-reduce jobs and other Hadoop functions are written in Pig, an open-source language designed specifically for Hadoop. Hive is an open source data warehouse originally developed by Facebook that allows for analytic modeling within Hadoop.

Hadoop: The Pros and Cons

The main benefit of Hadoop is that it allows enterprises to process and analyze large volumes of unstructured and semi-structured data, heretofore inaccessible to them, in a cost- and time-effective manner. Because Hadoop clusters can scale to petabytes and even exabytes of data, enterprises no longer must rely on sample data sets but can process and analyze all relevant data. Data scientists can apply an iterative approach to analysis, continually refining and testing queries to uncover previously unknown insights. It is also inexpensive to get started with Hadoop. Developers can download the Apache Hadoop distribution for free and begin experimenting with Hadoop in less than a day.

The downside to Hadoop and its myriad components is that they are immature and still developing. As with any young technology, implementing and managing Hadoop clusters and performing advanced analytics on large volumes of unstructured data requires significant expertise, skill, and training. Unfortunately, there is currently a dearth of Hadoop developers and data scientists available, making it impractical for many enterprises to maintain and take advantage of complex Hadoop clusters. Further, as the community improves upon Hadoop's myriad components and new components are created, there is, as with any immature open source technology/approach, a risk of forking. Finally, Hadoop is a batch-oriented framework, meaning it does not support real-time data processing and analysis.

The Hadoop Suitability Test

Hadoop ecosystem plays an integral part in any big data implementation. Despite the preference, not all enterprise use cases necessitate Hadoop as a must. How can we objectively assess the suitability of Hadoop to a business problem?

Below we have outlined few guiding principles to help in assessing the appropriateness of a Hadoop implementation with respect to the business problem.

- **Data Volume Consideration:** Historical as well Incremental in a scale of GB, TB, PB, EB

- **Data Type Consideration:** structured, semi-structured, unstructured

- **Data Integration and Interaction Mode Consideration:** batch, near real-time, real-time

- **Data Ingestion Pattern Consideration:** streaming, non-event-driven

- **Data Design Consideration:** local, distributed, centralized

- **Data Modeling Consideration:** ER, normalized, de-normalized

- **Data Access and Data Manipulation Consideration:** SQL, NoSQL

- **Workloads Consideration:** transactional, analytical

- **Data Store Design Considerations:** consistency, availability, partition-tolerance

- **Data Design Principle Consideration:** ACID, CAP, BASE

- **Contextualization/Association Consideration:** inter-element relationship, semantics

- **Compression Consideration:** ratio, performance overheads

- **Serialization Consideration:** Read Only, Write Only, Read/Write balance

- **Data Content/Analysis Type Consideration:** Key-Value Pairs, Document-Oriented, Graph-Centric

- **Latency Consideration:** low, medium, high

- **Network Performance Consideration:** memory, I/O, CPU, network

- **Security Consideration:** regulatory, access control, compliance, privacy

- **Data Platform Hosting Consideration:** physical, virtual, private/public/hybrid cloud

- **Data Quality Consideration:** High Quality, Mostly low focus on quality

- **Organization Adoption Maturity Consideration:** chasm, early taker, entrenching, mainstream, laggard, obsoleting

- **Big Data Product Support Consideration:** commercial, vendor, community, forum, broker, standards, practices

- **Skill Set Consideration:** competency, training, retooling, constraints, resources, tools

The Hadoop suitability test will help you assess your business problem mapped to all these parameters and a recommendation can be the drawn out of the test concerning whether you should go for Hadoop or not.

Additional Considerations for Big Data Warehouse (BDW)

The enterprise data platform must absolutely stay relevant to the business. As the value and the visibility of big data analytics grow, the enterprise data platform must encompass the new culture, skills, techniques, and systems required for big data analytics.

Sandboxes

BDW provides interesting capabilities to do exploratory analysis and experimentation. These capabilities usually consist of mashed up data sets, sophisticated algorithms, and codebase and rich data visualization components. We call these capabilities "sandboxes." Data analysts analyze the mashed up data sets with a wide variety of tools (mostly open-source tools to keep the cost low): data integration tools like the Hadoop ecosystem, sophisticated statistical analysis tools like *SAS, Matlab or R,* and many forms of ad hoc querying and rich data visualization tools like *Qlikview, Tableau.* Since BDW is an exploratory ground and aids in the discovery process, the data analyst responsible for a given sandbox has a complete freedom to do anything with the data (many times the data sources are well beyond the corporate firewalls) using any tool (often times the data analysts creates custom tools) to maximize productivity and enhance the discovery process. The sandbox capability has enormous potential but at the same time it also carries a significant risk of proliferation of isolated and incompatible stovepipes of data.

Exploratory sandboxes usually have lifetime association with a specific discovery process and objective. For example, the data analyst may be developing predictive models for a specific business hypothesis. Typically, if such an experiment produces a successful result, the sandbox experiment has met its goal, and the entire experimentation process along with data sets and algorithms are carefully evaluated to become a standard production feature. The data analyst then moves on to solve another problem.

Low latency

Many big data use cases are associated with real-time data processing, analysis, and in-sight generation. Low latency data processing and analysis needs are arising from the fact that data has a time dimension associated with it: if you do not process and analyze at that very moment, the value of data erodes significantly. An ideal implementation of low latency data processing and analysis would allow streaming data analysis to take place while the data is being acquired and processed. The availability of extremely frequent and extremely detailed event measurements can drive interactive intervention. The use cases where this intervention is important spans many situations ranging from online gaming to product offer suggestions to financial account fraud responses to the stability of networks.

Contextualizing the data

A key activity during discovery process is to build layers and layers of context associated with data. As you add more and more types of data, the mashed up datasets form a multi-layered interpretation engine for the original dataset. For example if a customer browses a website extensively before making a purchase, a great deal of micro-context is stored in all the webpage events prior to the purchase. When the purchase is made, some of that micro-context suddenly becomes much more important. These micro-contexts are pretty much meaningless before the purchase event, because there can be many activities on the webpage that are irrelevant events and may be inconsequential for analysis. However, once the purchase is made, if you have a long trail of all these micro-contexts captured, you can then reconstruct the sequence of events leading to a successful purchase. You can then apply this model to other ongoing web activities by customers and determine likelihood of purchases or by clever interventions you can influence the customer to make a purchase.

To Sample or Not to Sample

Exposing complete data sets (however big it may be) to a simple algorithm gives better results than exposing a sample of the data sets to a sophisticated algorithm. Interesting insights can be derived from very small populations within a larger data set that could be missed by only sampling some of the data. The analytics community is divided in their opinion about these two conflicting views.

Suppose you have a certain amount of data, and you look for events of a certain type within that data. You can expect events of this type to occur, even if the data is completely random, and the number of occurrences of these events will grow as the size of the data grows. These occurrences are "bogus," in the sense that they have no cause other than that random data will always have some number of unusual features that look significant but aren't. A theorem of statistics, known as the *Bonferroni* principle gives a statistically sound way to avoid most of these bogus positive responses to a search through the data.

Bonferroni's principle helps us avoid treating random occurrences as if they were real. Calculate the expected number of occurrences of the events you are looking for, on the assumption that data is random. If this number is significantly larger than the number of real instances you hope to find, then you must expect almost anything you find to be bogus, i.e., a statistical artifact rather than evidence of what you are looking for. This observation is the informal statement of Bonferroni's principle.

Big Data and Master Data Management (MDM)

In big data world, the data itself belongs to four different forms: data at rest, data in motion, data in many forms, and data in doubt. In addition, there are three styles of data integration prevalent in any enterprise scenario: bulk data movement, real-time, and federation.

Bulk data integration involves the extraction, transformation, and loading of data from multiple sources to one or more target databases. One of the key capabilities of bulk integration is extreme performance and parallel processing. Batch windows continue to shrink and data volumes continue to grow putting more stress on batch integration performance. Real-time integration involves low-latency integration and is often used in conjunction with complex event processing to enable real-time reporting and analysis. Federation is a completely different approach: it makes use of data through federated queries.

These three styles of integration should not be independent from one another. They should share a common foundation that establishes consistency in data. The process should be governed by enterprise data management principles such as data profiling, data quality assurance, improving the accuracy and completeness of data, tracking its lineage, and exposing enterprise metadata to facilitate integration. By applying a common approach to all three styles of integrations you can build a common foundation for information trust with common rules for data quality, metadata, lineage, and governance.

The data integration styles discussed above and the data characteristics go hand in hand in any enterprise data management scenario. For example, supplying trusted information to a data warehouse will require bulk data integration; but for specific reporting needs it may also need real-time integration, and potentially even federation to access other data sources. Building and managing a single view with MDM will again require bulk integration to populate MDM, real-time integration both to and from the MDM system, and federation to augment MDM's business services to blend data stored within MDM and data stored in other source systems.

While *master data management* approaches and implementation best practices have been around for some time, implications of MDM on big data platforms is relatively new. Big data is characterized by massive volumes, its high frequency, the variety of less structured data sources such as e-mail, sensors, smart meters, social networks, and weblogs, and the need to analyze vast amounts of data to determine value to improve upon management decisions.

Is MDM ready for Big Data Platforms?

A pertinent question always comes up: is MDM ready for big data? This question needs to be understood in the context of storage as well. In the traditional MDM implementations, you will see a MDM repository storing the master entities and operating under the defined MDM governance processes. In the traditional implementation approach, MDM is meant to be an operational, structured repository of key enterprise data entities: customers, households, products, locations, and many others.

However, in big data scenario, MDM isn't meant to be a big data repository, as it will never be able to store all social media data, transactional data, behavior data, etc. In big data scenarios, it is already evident and there will be more and more use cases that require MDM to integrate with variety of data sources that are not clearly defined.

Applying master data management to big data may be less about MDM and more about a paradigm shift in how we think about and use MDM. Although there are different ways to approach MDM, it's often seen as a repository for master data. All the data is dumped into MDM for sorting, cleansing, and achieving that mythical version of the truth.

But with big data the traditional thoughts need to change, as big data is too big and changes too fast. Historically, MDM was around the customer repository. That's just not feasible in the world of Facebook and Twitter. Information is federated, so the next generation of MDM designers will need to think less about MDM as a repository and more about MDM as a way to govern global information. MDM programs will ultimately need to govern the relationships between internal data and big data from external sources.

Obviously, MDM done well has always had governance as a key component, because a good master data management program will cover who owns data and who has the authority to alter it; otherwise, the data is fixed briefly and then becomes outdated or the conflicts reappear. But big data puts even more demands on MDM as a governance tool, because there's so much data, the focus shifts from just adding it to filtering out what's usable and useful to the business.

The forthcoming generations of MDM tools need to have characteristics of a service-level agreement infrastructure, with MDM offering a cross-reference of data and control of the core that matters. MDM will need a data integration infrastructure to create and share that whole view as needed. It is more about shifting the information when it's required and providing that identification and less concerned with the historical view of effectively a digital landfill of data that everybody poured everything into.

Traditional approaches to master data have led us to think of it as a single data entity; master data is all about the linked data elements for a single record, and no duplication or variation should ever exist, thus ensuring consistency and uniqueness. Master data in the current thinking represents a defined, named entity (customer, supplier, product, etc.). This approach is tied to an application (customer resource management, enterprise resource management) for a particular business unit (marketing, finance, product management, etc.). It may have been the entry point for MDM initiatives, but it's difficult to expand that master data to other processes, analysis, and distribution points. Master data as a static entity only takes you so far, regardless of whether big data is incorporated into the discussion or not. This is a very static view of master data.

Data that matters have always been represented by *what, why,* and *when.* Big data introduces another interesting and critical characteristic to equation: the "who." In essence, *who* represents and brings out the context associated with data elements. Let's take a look at customer master data. In this context, big data is interesting because it provides an understanding of what drives behavior. In most cases, this means shifting priority for data quality to transactional data and metadata over master data. Master data thus is expanded to classify the behavior, time, and intent domains.

What this means is that you move from a two-dimensional model to a multidimensional model of master data. Master data is all about the data model both in terms of relationships and hierarchies and how data elements are combined. Master data, metadata, and reference data converge under an MDM umbrella, allowing for unlimited combinations determined by categories, definitions, and context. Because data has moved beyond structured and relational database constraints due to big data characteristics, MDM must account for the structure and enforce business policies for a trusted holistic view. Thus, in a big data world, the approach to master data model must go beyond and over the uniqueness of the data entity and should include other relevant dimensions.

In order for MDM to work with big data systems, there are several requirements:

- MDM must also be able to store profiles for any big data source that needs to be linked to a master record. Examples include: account IDs to link transactional big data to customer and account records, mobile device IDs to link mobile device data, and real-time location data to a customer record, among others.

- The MDM system must also be able to store preferences for each big data source. *Does a customer want you to analyze their tweets? Or their Facebook profile?* MDM must track the customer's preferences and consent for certain types of communication and interaction.

- MDM should relate many-to-many relationships between customers and profiles. For example, a household is related to a single social media profile on a photo-sharing website (one social media profile for many customers who belong to a household). This enables MDM to effectively feed a big data application with relevant master data and big data links.

- MDM must also be able to store the output from big data analytics. Intent to purchase, next best action, customer churn alert flags, negative customer sentiment: these are all attributes that should be stored in MDM. Insights from big data should be available to multiple operational channels (for example, if you detect that a customer is dissatisfied with your company, then you want all the interaction channels to know that fact, no matter which channel the customer interacts with).

- The MDM system should also have the capability to proactively detect events and send event notifications, triggering action in business applications and enterprise processes as necessary. MDM must be an active participant in big data analytics.

- The big data system must be able to interact with MDM. Whether you're working with transactional data, analyzing social media data, or analyzing streaming call detail data off a network, the big data system needs to understand the master view of customers and products. There's no point in the big data system re-inventing the wheel and trying to determine unique records and identities. This integration or information exchange aspect is important from a discovery/experimentation type of workload point of view as well, which is the most-cited big data system usage so far. In essence, big data applications need to be MDM-aware. They should obtain master data from MDM either in batch load or in real-time if necessary.

- MDM systems must be capable of integration with big data systems via batch and possibly real-time SOA as required. There should be a bi-directional relationship between MDM and big data; big data technology can feed insights to MDM, and MDM can feed master data definitions to big data. In some ways, this is similar to the relationship between MDM and the data warehouse: MDM will both receive and feed that system.

We have already seen a number of use cases of MDM and big data working together: e.g., social media analytics to predict customer churn or intent to purchase, mobile network analytics to make real-time location-specific product offers, and multi-channel interaction analysis to predict and prevent customer churn. However, there are pitfalls too.

Consider the use case, "employ big data technology to mine social media and understand intent which would unearth potential new customers." But what if those same prospects were already customers? And what if your CRM systems already knew, or should have known, the prospects intentions? In order to make a targeted and purposeful analysis of big data, you need a starting point, and that starting point should be, understanding your existing customers through MDM.

This highlights the first aspect of the big data and MDM. MDM feeds big data. MDM can provide master definitions of customer, household, relationship, and product hierarchies to big data. When your requirement moves from aggregate analysis (e.g., general market sentiment toward your company) to specific analysis (e.g., which customers have an intent to purchase product X), that is when you require master data to guide big data analysis.

Big data technology can process and analyze unstructured data sources (e.g., PDF documents) to determine unique identities and relationships among master data entities. This technology can also analyze third-party data (unstructured PDF documents on company financials and ownership) to help determine organization parties and hierarchies. You could potentially accelerate your initial MDM implementations by extracting master data from previously untapped big data sources. For example, you can analyze SEC filing documents for risk exposure, to understand customers, their financial health, and key individuals at those companies. The danger in big data projects lies in not recognizing the requirement for MDM and treating data quality, matching, and storing unique records as a one-off tactical task.

Alternatively, you can start doing a search index for big data. Start with already defined enterprise master data entities and then analyze new sources of data for specific master data records. Don't analyze all customers, analyze the most valuable ones. Don't analyze all of your products, analyze the most profitable ones. This may initially be expressed entirely as an analytics requirement from business owners. Consider the use case, "analyze social media to understand potential online bets the customers might make." What does this mean? Who are these customers? What constitutes a "betting event"? And how will the company respond in time to capture that opportunity?

At the most fundamental level the company needs to have the answer to the question – who are their customers? There's no point in analyzing all available social media feeds and then determining who your customers are. There are 2 billion Internet users globally. How many customers do you have? Less than 2 billion? Doesn't it make sense to start the other way around? Know what you're looking for before you start looking.

MDM and Big Data Integration Scenarios

Are there any guidelines and best practices available? Below we will discuss a few scenarios to give you a head start on how to integrate MDM and big data:

- **You are searching and matching for the same entity types over and over.** If your big data project requires you to know whether a social media blogger is a customer, and you will run this same determination every time an interesting social media post is detected, then you have a master data problem. You need to know your customers.

- **You are performing targeted analysis, not an aggregate analysis.** When you are looking for particular product feedback to respond to isolated incidents versus general sentiment toward your brand, or you are looking for a particular customer's multi-channel service experiences versus tracking the general service levels, then you have a master data problem. You need to know specific customers and products in order to guide your big data analysis.

- **You want to combine the analysis of multiple master data domains from new big data sources.** If your big data use case involves matching multiple data domains and deriving new insights from big data sources, you likely have an MDM requirement. For example, telecommunications companies are increasingly interested in mobility: i.e., understanding the location of mobile devices and the potential opportunities (selling new products, proactive service alerts, etc.). In order to realize this use case, the company will need to understand unique accounts, devices, customers, households, and locations. This is a multi-domain MDM problem to start with.

MDM Hub as a Foundation for Big Data

Master data is really just a subset of big data. This subset tends to be already in structured format, reasonably trustworthy, and shared and common across different lines of business or departments.

When clients start discussing big data projects, they usually want to start making better use of all their data beyond just the core master data elements. As organizations plan these projects, it is crucial that they leverage their MDM hub as a foundation for big data.

The MDM hub is where you keep the most complete view of your customers, products, accounts, and more. As you uncover more information about those same entities, the MDM hub is the logical place to keep those new insights. The MDM hub can keep a traditional "golden record" of trusted information side-by-side with a less-trusted view of the same person or product based on what you find among your big data. These two views can be combined to provide a more insightful complete view, but they can still be kept separately in cases where your business can't afford to base decisions on the less-trusted view. The MDM hub can already tell you who your customers and prospects are, so use that knowledge to more efficiently sift through the rest of your big data to find more about those same customers and prospects.

■ **Note** when you combine the traditional "golden record" with new information found among your big data, the superset of information can power even better business insights and business decisions that were not possible before.

Big data discussions have been gaining momentum and substance, but not much distance has been covered when it comes to master data management with respect to big data implications. The value proposition is to bring MDM into big data analytics to further enrich the master entities. For example, the amount of comments that are collected and collated by product-marketing teams is humongous. These comments can be found across the Internet in discussion forums, personal blogs, and other places. All this data, however, follows a typical big data pattern. It is large and builds up quickly; it is semi-structured and comes in fast and furious, posing a challenge to extract, transform, and load it to relational databases. In order for all this information to be useful, free-form comments need to be tied in with product catalogs. There is potentially important information in all this consumer feedback and the product forms the central point of intersection of MDM and big data.

In summary, Figure 5-7 shows the various interactions that a multi-domain MDM system should have with respect to big data use cases.

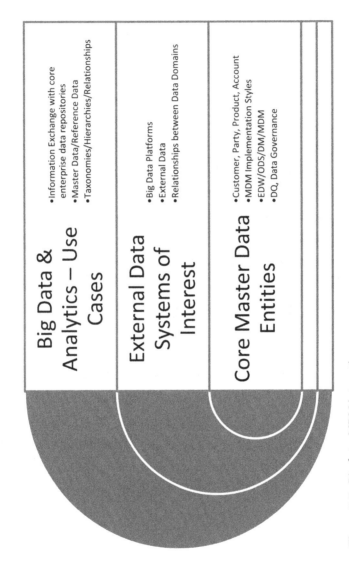

Figure 5-7. *Big data: MDM interactions*

Big data represents many things. For some, it's the technology we use to store, retrieve, and manage petabytes of data that we create each day. For many in enterprise technology, big data means using sophisticated techniques to analyze the large and growing volumes of transaction data in order to improve business decision making.

In big data and analytics scenarios, instead of being constrained to just intra-domain customer hierarchy dimensions following traditional MDM implementation styles, we will need a multi-domain MDM approach to view transactional data and external data entities through different data domain relationships (as shown in Figure 5-8). Sales could analyze data by customer, territory, and geography. Marketing could see the evolution of the buying process by campaigns, social media interactions, and click- stream analysis; finance can get a view across suppliers and sales, and marketing can also get an insight into the brand reputation and sentiments by looking at external data.

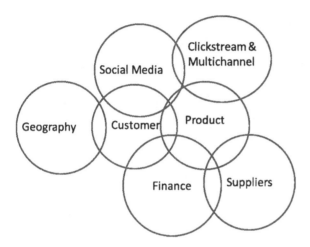

Figure 5-8. *Big data: MDM multi-domain interaction*

Figure 5-9 illustrates a big data tool: MDM integration logical architecture. The focus of this architecture is to illustrate the various components in an enterprise data management landscape and how they interact or leverage MDM implementation integrating with business systems across the enterprise including big data platforms. The master data management services and information integration services are fundamental to multiple master data domains such as product, customer, supplier, account, and location.

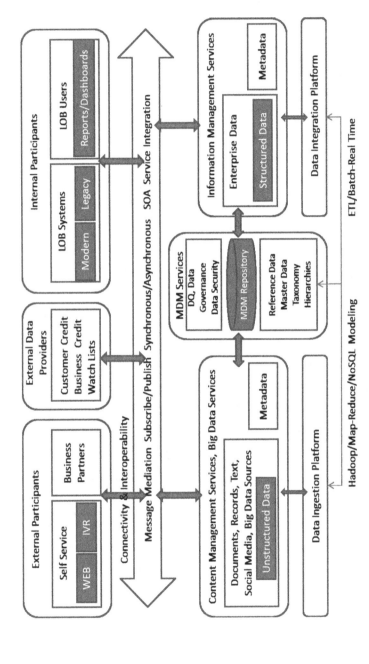

Figure 5-9. Big data: MDM integration logical architecture

- *External participants* may access and update master data through multiple interaction channels. Customers might access and update master data through business systems that provide self-service capabilities for shopping and online channels or through the use of telephony systems to access and update personal information. Supply-chain data from suppliers, trading partners, and business partners participate in business-to-business transactions that involve the exchange of core master data entities such as customer and product data. Agents that conduct business on behalf of a company may access and update master data through a business system provided by that company or through a business-to-business transaction. Business system users update and query master data typically through the use of their respective business systems.

- Data from *external data providers* such as as *Dun and Bradstreet, Acxiom, Lexis Nexis, Ac Neilson, Fair Issac,* and *Credit Bureaus* can be utilized for additional information about a person or organization to enrich master data maintained in the MDM System. Data from these sources may be used to support the initial loading of master data into the MDM system or periodic updates, or data may be used on a transactional basis based upon business requirements. Government agencies also provide watch lists required to support regulatory compliance.

- The *connectivity and interoperability layer* serves as an enterprise information exchange backbone connecting business-to-business interactions with partners, system-to-system interactions within the enterprise, and interactions with the external data providers. Instead of having many point-to-point interfaces between systems, this layer creates one single interface using application integration techniques such as enterprise application integration hubs that support communications through the use of messaging, or have adopted the use of an enterprise service bus. This layer also facilitates communications between MDM services, information management services and content management services, and big data services. The connectivity and interoperability layer represents the enterprise service bus architectural construct, or it can simply be thought of as a layer that provides choreography services and synchronous and asynchronous integration capabilities such as message mediation and routing, publish and subscribe, FTP, and service-oriented integration through the use of web services.

- The *master data management services* component consists of a set of services that are grouped into the following components:
  - *Interface Services* support a consistent entry point to request MDM services through techniques such as messaging, method calls, web services, and batch processing. The same

MDM service should be invoked during batch processing, which may be requested as part of a transaction in order to maintain and apply consistent business logic.

- *Lifecycle Management Services* manage the lifecycle of master data, provide *CRUD* (create, read, update, and delete) support for master data managed by the MDM system, and apply business logic based upon the context of that data. Lifecycle management services call data quality management services to enforce data quality rules and perform data cleansing, standardization, and reconciliation. Event management services are called to detect any actions that should be triggered based upon business rules or data governance policies.

- *Hierarchy and Relationship Management Services* manage master data hierarchies, groupings, and relationships that have been defined for master data. These services may also request identity analytics services to discover relationships, such as those between people that are not obvious, and then store that information in the MDM system.

- *Master Data Management Event Management Services* are used to make information actionable and trigger operations based upon events detected within the data. Events can be defined to support data governance policies, such as managing changes to critical data, based upon business rules or time and date scheduled.

- *Authoring Services* provide services to author, approve, manage, customize, and extend the definition of master data as well as the ability to add or modify instance master data, such as product, vendor, and supplier. These services support the MDM collaborative style of use and may be invoked as part of a collaborative workflow to complete the creation, updating, and approval of the information for definition or instance master data.

- *Data Quality Management Services* validate and enforce data quality rules, perform data standardization for both data values and structures, and perform data reconciliation. Data quality management services also include data profiling, analysis, cleansing, data standardization, and matching services. Data profiling and analysis services are critical for understanding the quality of master data across enterprise systems and for defining data validation, data cleansing, matching, and standardization logic required to improve master data quality and consistency.

- The *Master Data Repository* consists of master data and metadata for the MDM system and history data that records changes to master data. Master data management services can also be used to maintain and control the distribution of reference data that should be maintained at the central level for an organization. As discussed in prior sections, the MDM data repository can also include the various inferences drawn during discovery and experimentation type of work done in the big data platforms.

- The *MDM services* components also enable information exchange between the content management services, big data services, and information management services, thereby providing a rich data services ecosystem that is consistent and integrated across the enterprise.

- *Information management services* provide ETL services primarily for batch and real-time integration of structured data, and EII services for federated query access to structured and unstructured data distributed over disparate data sources. ETL services support the initial and incremental extract, transform, and load of data from one or more source systems to meet the needs of one or more targets, such as a data warehouse and MDM system.

- The *content management services and big data services* provide mechanisms to capture, aggregate, and manage unstructured content in a variety of formats such as images, text documents, web pages, spread sheets, presentations, graphics, e-mail, video, and other multimedia. The Hadoop and map-reduce technologies provide the ability to search, catalogue, secure, manage, and store unstructured content and workflow services to support the creation, revision, approval, and publishing of content. In conjunction with the MDM data repository and the metadata contained therein, now you can have the ability to identify new categories of content and create taxonomies for classifying enterprise content. Other aspects like records management and storage management are part of the content management and big data services. Records management services include management of the retention, accessing control and security, auditing and reporting, and ultimate disposition of business records. Storage management services provide for the policy-driven movement of content throughout the storage lifecycle and the ability to map content to the storage media type based on the overall value of the content and context of the business content.

The discovery and experimentation type of workloads that uses a combination of data platforms and associated services can effectively leverage the MDM services components to discover non-obvious relationships between various data entities such as those that are part of the same household but have different name and address information and between people and organizations. The taxonomy hierarchy and relationship information between data entities can be effectively leveraged by big data platforms while analyzing data from external data sources along with data from within the enterprise.

The MDM logical integration architecture is designed to support the multiple MDM methods of use across multiple master data domains, to maintain cross-domain relationships, and to provide the required functionality to have a collaborative environment taking into account the hybrid architectures of relational and non-relational data platforms. The architecture is structured to be scalable, highly available, and extensible, and provides the flexibility to integrate technology from a variety of vendors and integrate with future unknown systems.

Data Quality Implications for Big Data

There is a lot of literature about what is now possible given the opportunity of big data and what organizations should be doing. But very little has been discussed in terms of guidance and recommendations related to data quality and big data.

Data management and data quality principles for big data are the same as they have been in the past for traditional data. But priorities may change, and certain data management and data quality processes such as metadata, data integration, data standardization, and data quality must be given increased emphasis. One major exception involves the time-tested practice of clearly defining the problem. In the world of big data, where data may be used in ways not originally intended, data elements need to be defined, organized, and created in a way that maximizes potential use and does not hinder future utility.

Your data quality approach for big data should be designed with several factors in mind: it doesn't make sense to apply one data quality approach for all types of data. You should consider where the data came from, how the data will be used, what are the workload types, who will use the data, and perhaps most importantly, what decisions will be made with the data.

What data do you trust? Increasingly, business stakeholders and data scientists are beginning to draw conclusions based on big data sources. Yet, the fact is, these data are mined and analyzed in a way that doesn't adhere to the existing data governance processes. There is a valid argument for doing it this way. If you need speed of insight and support data discovery over repeatable reporting, then you can't constrain the activities.

Traditional approaches to data quality heavily revolve around the notion of persistence of cleansed data. For years data quality efforts have focused on finding and correcting bad data. We use the word *cleansing* to represent the removal of what we don't want. Knowing what your data is, what it should look like, and how to transform it into submission defined the data quality handbook. Whole practices were created to track data quality issues, establish workflows and teams to clean the data, and then reports were produced to show what was done. These practices were measured against metrics such as identification of the number of duplicates, completeness of records, accuracy of records, currency of records, and conformance to standards, to name a few. However, when it comes to big data, how do we cleanse it?

The answer to the above question is, maybe you don't. The nature of big data doesn't allow itself to traditional data quality practices. The volume may be too large for processing. The volatility and velocity of data makes it difficult to keep track of. The variety of data, both in scale and visibility, is ambiguous.

Running faster won't get you to the right place if you don't know where you're going. By creating better, faster, and more robust means of accessing and analyzing large data sets can lead to erroneous outcomes if your data management and data quality processes don't keep pace.

Traditionally, whenever data quality concerns were raised we always asked the following questions:

- What are the data quality benchmarks to ensure the data will be fit for its intended use?

- What are the key data qualities attributes to be measured (for example, validity, accuracy, timeliness, reasonableness, completeness)?

- What approaches will we take to manage data quality? For example, should we fix issues at the source or have a cleansed and quality assured environment downstream?

- How do we capture the data lineage and traceability (for example, data flows from the underlying business processes) aspects?

Will these traditional methods be relevant for big data scenarios, or we will need new principles and processes? What are the data management and data quality implications of these technologies?

Let us discuss few of the critical aspects related to the data life cycle and big data implications that heavily influence data quality.

Metadata. Metadata is important to any data management activity. Metadata and metadata management become even more important when dealing with large, complex, and often multi-sourced data sets. Metadata to be used across the enterprise must be clear and easily interpreted and must apply at a very basic level.

Data Element Classification. For big data quality and management (big DQ and DM), minimum metadata requirements need to be established and, ultimately, metadata standards too. To foster cross-enterprise use of data, taxonomies (classification or categorical structures) need to be defined, such as demographic data, financial data, geographic/geospatial data, property characteristics, and personal identifiable information.

Data Acquisition. While acquiring data, it is critical for data to be organized to be more readily assessable. Data exchange standards for big DQ and DM are key aspects in the acquisition process. Use of the common vocabulary and definitions facilitates the mapping of data across sources.

Data Ingestion and Integration. Integrating data across multiple sources is certainly a large part of a big data effort. One school of thought is to create a "data lake" where you dump data coming from various sources, and then later on as you start using the data, you define standards, establish lineage, and create metadata definitions. While this approach significantly reduces the process-related bottlenecks, it also creates concerns around quality of data. Usage of tools and processes like MDM, entity resolution, and identity management will surely help to address some of the data- quality-related concerns.

While data quality has traditionally been measured in relation to its intended use, for big data projects, data quality may have to be assessed beyond its intended use and one may have to address how data can be repurposed. To do so, data quality attributes—validity, accuracy, timeliness, reasonableness, completeness, and so forth—must be clearly defined, measured, recorded, and made available to end users.

Artifacts relating to each data element, including business rules and value mappings, must also be recorded. If data is mapped or cleansed, care must be taken not to lose the original values. Data element profiles must be created. The profiles should record the completeness of every record. Because data may migrate across systems, controls and reconciliation criteria need to be created and recorded to ensure that data sets accurately reflect the data at the point of acquisition and that no data was lost or duplicated in the process.

Special care must be given to unstructured and semi-structured data because data quality attributes and artifacts may not be easily or readily defined. If structured data is created from unstructured and semi-structured data, the creation process must also be documented and any of the previously noted data quality processes applied.

In a big data scenario, you must create data-quality metadata that includes data quality attributes, measures, business rules, mappings, cleansing routines, data element profiles, and controls.

High Availability versus High Data Quality

Typically, big data solutions are designed to ensure high availability. High availability is based on the concept that it is more important to collect and store data transactions than it is to determine the uniqueness or accuracy of the transaction. Some common examples of big data/high availability solutions are Twitter and Facebook.

It is possible to configure a big data solution to validate uniqueness and accuracy. However, in order to do so you need to sacrifice some of the aspects of high availability. So, in some regard, big data and data quality are at odds.

This is because one of the fundamental aspects of high availability is to write transactions to whichever node is available. In this model, consistency of transactional data is sacrificed in the name of data capture. Most often, consistency is eventually configured for queries or on data reads as opposed to data writes.

In other words, at some given point in time you do not have consistency in a big data set. Even more troubling is the fact that most transactional conflicts are resolved based on timestamps. This is to say that the most recently updated transaction is commonly regarded as the most accurate. This approach is, obviously, an issue that requires further examination.

Why we don't see an inherent trade-off between the volume of a data set and the quality of the data maintained within it?

We are under the mistaken impression that there's an inherent trade-off between the volume of a data set and the quality of the data maintained within it. In essence, big data sets are big, and hence it is natural to deduce that there is a good amount of inconsistent, inaccurate, redundant, out of date, or un-conformed junk data. This way of thinking may have some merit; however, let's understand the reality. When you talk about big data, you're usually talking about more volume, more velocity, and more variety. Of course, that means you're also likely to see more low-quality data records than in smaller data sets.

But that's simply a matter of the greater scale of big data sets, rather than a higher incidence of quality problems. While it is true that a 1 percent data quality issue is numerically far worse at 1 billion records as opposed to 1 million, the overall percentage remains the same, and its impact on the resulting analytics is consistent. Under such circumstances, dealing with the data cleanup may require more effort—but as we noted earlier, that's exactly the sort of workload scaling where big data platforms excel.

Big data isn't the transactional source of most data problems. The cause of data quality problems in most organizations is usually at the source transactional systems—whether that's your customer relationship management (CRM) system, general ledger application, or something else. These systems are usually in the terabyte range. Any situation where you fail to keep the system of record cleansed, current, and consistent, you can expect the magnitude of data-quality-related issues. If you can't fix the issues at the source systems, you can take an alternative approach to fix the issues downstream (through EDWs and MDM implementations) by aggregating, matching, merging, and cleansing data in intermediary staging databases.

The quality problem has everything to do with inadequate controls at the data's transactional source, and very little to do with the sheer volume of it.

Big data is about aggregating new data sources that you haven't historically needed to cleanse. In data warehouse systems the issue of data quality is fairly well understood: you are primarily concerned with maintaining the core systems of record such as customers, finances, human resources, the supply chain, and so on. In contrast, a lot of big data initiatives are for deep analysis of aggregated data sources such as social marketing intelligence, real-time sensor data feeds, data pulled from external resources, browser click-stream sessions, IT system logs, and the like. These sources have historically not been linked to official reference data from transactional systems. From an enterprise data management processes perspective, there was clearly no focus, and these data sources needed to be cleaned because they were looked at in isolation by specialist teams that often worked through issues offline and weren't feeding their results into an official system of record. However, cross-information-type analytics—which is common in the big data space—have changed this dynamic.

Although individual data points can be of marginal value in isolation, they can be quite useful when pieced into a larger puzzle. They help provide context for what happened, or what is happening.

Unlike business reference data, these new sources do not provide the sort of data that you would load directly into your enterprise data warehouse. Rather, you drill into it to distill key patterns, trends, and root causes, and you would probably purge most of it once it has served its core tactical purpose. This generally takes a fair amount of mining, slicing, and dicing.

Data quality matters in two ways in this situation. First, you can't lose the source, inferences are drawn from the data, and actions are taken while distilling the data—and these items need to be defined consistently with the rest of your data. Second, you can't lose the lineage of how you performed the analysis.

The, who, what, when, where, and how need to be discoverable and reproducible.

Keep in mind that often when we're talking about big data we are talking about using data that we haven't been able to exploit well in the past—so we're typically trying to solve different problems. We're not trying to figure out the profitability of each of our stores. We should already be doing that using high-quality data from systems of record and doing the things we do to standardize and reshape as we put it into a data warehouse. What we're trying to do here is find out what's contributing to the profitability for the stores.

Big data allows you to find quality problems in the source data that were previously invisible. If you're aggregating data sets into your big data platform that have never coexisted in the enterprise data ecosystem before, and if you're trying to build a common view across them, you may be in for a rude awakening. It's not uncommon to find quality issues when you start working with information sources that have historically been underutilized.

When looking at underutilized data, quality issues can take you through nasty discoveries, so it pays to expect the unexpected. For example, in many cases you may find that the system data provided as a reference is highly variable and not as described in the specifications. In cases like this, you either need to go back and deal with the core system data generation process or work past the quality issues. This is a fairly common occurrence since, by definition, when you are dealing with underutilized information sources, this may be the first time they have been put to rigorous use.

This issue rises to a new level of complexity when you're combining structured data with unstructured sources that—it almost goes without saying—are rarely managed as official systems of record. In fact, when dealing with unstructured information (which is the most important new source of big data), expect the data to be fuzzy, inconsistent, and noisy. A growing range of big data sources provide non-transactional data—event, geo-spatial, behavioral, click stream, social, sensor, and so on—that is fuzzy and noisy by its very nature. Establishing a corporate standard and shared method for processing this data through a single system is a very good idea.

Interestingly, big data is ideally suited to resolve one of the data quality issues that has long impacted the statistical analyses: the traditional need to build models on training samples rather than on the entire population of data records. This idea is important but under-appreciated. The scalability constraints of analytic data platforms have historically forced modelers to give up granularity in the data set in order to speed up model building, execution, and scoring. Not having the complete data population at your disposal means that you may completely overlook outlier records and, as a result, risk skewing your analysis only to the records that survived the cut.

This isn't a data quality problem (the data in the source and in the sample may be perfectly accurate and up to date) as much as a loss of data resolution downstream when you knowingly filter out the sparse/outlier records. Let's look at a specific example in the messy social listening space. It's easy to manage noisy or bad data when you are dealing with general discussion about a topic. The volume of activity here usually takes care of outliers, and you are—by definition—listening to customers. Data comes from many sources so you can probably trust (but verify through sensitivity analysis) that missing or bad data won't cause a misinterpretation of what people mean. However, when you examine what a particular customer is saying and then decide how you should respond to that individual, missing or bad data becomes much more problematic. It may or may not be terminal in that analytics run, but it inherently presents more of a challenge. You need to know the impact of getting it wrong and design accordingly.

Your data quality efforts need to be defined more as profiling and standards versus cleansing. This is better aligned to how big data is managed and processed. While on the surface, big data processing is batch in nature, it would seem obvious to institute data quality rules the way they have always been done. But the answer is to be more service-oriented, invoking data quality rules that provide improved standardization and sourcing during processing versus fundamentally changing the data. In addition, data quality rules are invoked in a customized fashion based on customer service calls from big data processing.

Why this also makes sense is that when you do decide to persist sourced big data into your internal infrastructure, you have pre-aligned the data to existing policies for integration and business rules for improved mapping and cleansing that would need to persist. In essence you treat big data as a reference source, not a primary source. So, think about data quality in the context of supporting preprocessing with Hadoop and map-reduce through profiling and standards, not cleansing.

In many cases, big data involves some form of textual or unstructured data. Quality issues that plague text from user-entered data largely applies to big data initiatives. The following examples represent typical data quality challenges relating to text that should be extended into big data environments:

- Identifying misspelled words or managing synonym lists for grouping similar items like "lvm," "left voice mail," "left a message," etc., that may affect analysis.

- Leveraging content categorization to ensure that the textual data is relevant. For example, filtering out noise in textual data relating to a company name: differentiating SAS Institute, SAS shoes, SAS the airline, etc.

- Utilizing contextual intelligence to discern meaning. For example differentiation between the person and the name of a hotel, "Paris Hilton walks into the Paris Hilton." This should include the ability to factor this into count or summary analysis where it is necessary to delineate between the person and place.

There are several other considerations for data quality for big data scenarios listed below:

Consider the type of data: The data quality requirements for different forms of data will vary and your approach should match the needs of the data. For example:

- Big data projects that relate to traditional forms of data like transaction data related to key entities like customers, products, etc., can leverage existing data quality processes as long as it scales to meet the needs of massive volume.

- Big data originating from machines or sensor data (e.g., RFID tags, manufacturing sensor data, telecom networks/switches, utilities, etc.) will not be prone to errors as compared to data that is entered by humans. As additional sensor data streams in you need to ascertain the difference between signals and noise: for example, a pigeon sitting on a sensor causing the data to throw up random alarms.

- Social media data such as Twitter, Facebook, etc., may look highly unstructured, but they still contain a structure around it: a meta-data description defining type of tweet stream and then the text string that contains the content of the tweet. From a data quality perspective, this will involve a combination of entity matching, monitoring to ensure that the tweet stream is not interrupted along with the ability to analyze the text.

Not all analysis requires exactness: If you are attempting to identify a general pattern and you have a lot of data, the volume of data is not likely to impact the overall conclusion. For example, if you have a massive amount of click stream data and you are looking for patterns (when people leave a site, which path is more likely to result in purchase or conversion, etc.) the outliers will not impact the overall conclusion. In this case, it's more of an analytics process versus a data quality process. However, you will still have to check the relevance aspects of data: for example, if someone accidentally ends up on your website, they aren't really part of the population that you are concerned with (unless you are analyzing why they are there in the first place). Same with bots versus actual users; bot traffic is not likely to be erroneous, but it is possible to extend your data quality efforts to include relevance as a quality.

Don't cleanse away analytical value: In the traditional analytics approach, data cleansing is a critical task, and the argument was fairly straightforward: you do not want to run your algorithms on data that is not cleansed, as it may negatively influence the outcomes. However, this cleansing approach should not be applied to all analytics cases. For example, if you are tracking risk outliers, unusual transactions should not be cleansed away because they fall outside of the norm and may represent fraud.

Processes up front in the analytical lifecycle like data discovery, data exploration, opportunity identification, data relationship identification, etc., are better performed on the data prior to any cleansing taking place. For example, assessing the value of the various attributes by analyzing access frequency, detecting outliers, or discovering correlations between attributes may form the initial stages in understanding data distribution. Then once it is clear about the questions you are driving toward, the type of analytics that will be leveraged, etc., you can make the proper determination about data quality, etc. You may even leverage a gradual cleansing process as part of your strategy.

Using analytics to assess quality: Use data quality processes to determine the impact of missing attributes or purposely fabricated data on analytic algorithms. Data quality or MDM processes can be effectively leveraged to correlate the big data source to the transaction or enterprise data. This allows you to relate specific customer feedback from social sources with internal customer data that is tied to product or service purchases. With contextual data (social media, sentiment analysis, opinion mining), these cross validations are valuable. For example, most of the social data is self-reported. People self-report about their shopping experience, their likes/dislikes, etc., and chances are this information is misrepresented by the user (they intentionally fabricate their experiences and opinions to develop a story among their friends). In this case, traditional data quality approaches will not be useful, but analytics can be used to provide some level of value assessment. Same with sentiment data, considering transactional data and sentiment data relating to purchase behavior: if sentiment is negative and purchase behavior is positive, this could indicate a data quality problem, or it could relate to the customer being locked in without additional choices. This approach can strengthen marketing analysis efforts since the analytics is correlated at the individual customer level versus correlating broad segments of transactional and interaction data. You can further extend to entity match friends from the actual customer to determine if the customer interaction drives business with the customer's friends.

Putting it all Together – A Conceptual BDW Architecture

In the sections above we discussed several architecture styles and implementation approaches. An enterprise data platform ecosystem consisting of BDW, EDW, DQ, MDM, and analytics can become mind-boggling. To help you understand the various components of a big data solution, Figure 5-10 is an attempt to put together a conceptual architectural view of a big data platform.

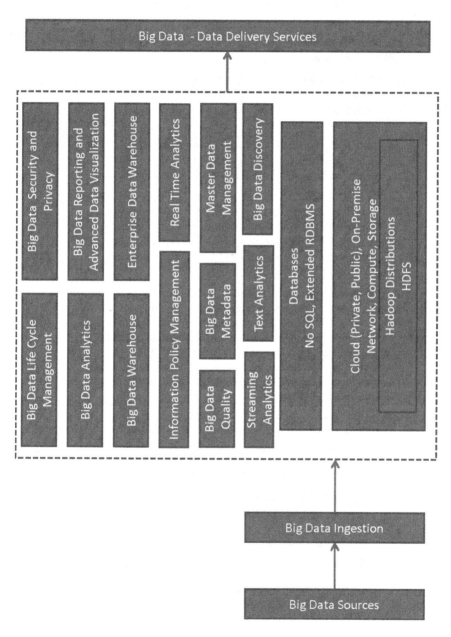

Figure 5-10. *Conceptual BDW architecture*

- **Big Data Sources.** Big data types include web and social media, machine-to-machine, big transaction data, biometrics, and human-generated data. This data may be in structured, unstructured, and semi-structured formats.

- **Big Data Ingestion.** Big data ingestion technologies fall into a few different categories:

  1. *Bulk data movement.* Bulk data movement includes technologies such as ETL that extract data from one or more data sources, transform the data, and load the data into a target database.

  2. *Data replication.* Replication technologies like change data capture can capture big data, such as utility smart meter readings, in near real time with minimal impact to system performance.

  3. *Data virtualization.* Data virtualization is also known as data federation. Data virtualization allows an application to issue SQL queries against a virtual view of data in heterogeneous sources such as in relational databases, XML documents, and on the mainframe.

- **Hadoop Distributions.** Hadoop distributions consist of a large number of technologies with their own release schedules. A number of vendors have created their own commercial distributions of Apache Hadoop that have undergone release testing and bundle product support and training. Most enterprises that have deployed Hadoop for commercial use have selected one of the Hadoop distributions: *Cloudera, MapR, Hortonworks.*

- **Databases.** Enterprises have the ability to select from multiple database approaches:

  1. NoSQL ("not only SQL") databases are a category of database management systems that do not use SQL as their primary query language. These databases may not require fixed table schemas and do not support join operations. These databases are optimized for highly scalable read-write operations rather than for consistency. NoSQL databases include a vast array of offerings such as Apache HBase, Apache Cassandra, MongoDB, Apache CouchDB, Couchbase, Riak, and Amazon DynamoDB. DataStax offers an enterprise edition that includes a Hadoop distribution, and replaces HDFS with the CassandraFS.

2. In-memory database management systems rely on main memory for data storage. Compared to traditional database management systems that store data to disk, in-memory databases are optimized for speed. In-memory databases will become increasingly important as organizations seek to process and analyze massive volumes of big data. SAP HANA, Oracle TimesTen In-Memory Database, and IBM solidDB are all examples of in-memory databases.

3. Apache Sqoop is a tool that allows bulk transfer of data between Hadoop and relational databases. In addition, software vendors are also upgrading their database offerings to co-exist with Hadoop ecosystem: the Oracle loader for Hadoop uses MapReduce jobs to create data sets that are optimized for loading and analytics within the Oracle relational databases. IBM InfoSphere BigInsights includes a set of Java-based user-defined functions (UDFs) that enable integration with IBM DB2 using SQL. Microsoft offers a bi-directional Hadoop connector for SQL Server.

4. Legacy database management systems rely on non-relational approaches to database management. Vendors will increasingly re-tool these systems to support big data. For example, the IBM DB2 Analytics Accelerator for z/OS leverages the *IBM Netezza* appliance to speed up queries issued against a mainframe-based data warehouse running IBM DB2 for z/OS.

- **Streaming Analytics.** Hadoop is well suited to handle large volumes of data at rest. However, big data also involves high velocity data in motion. Streaming analytics, also known as *complex event processing* (CEP), refers to a class of technologies that leverage massively parallel processing capabilities to analyze data in motion as opposed to landing large volumes of data to disk. There are a number of open-source and vendor tools in this space. For example, Apache Flume is an incubator effort that uses streaming data flows to collect, aggregate, and move large volumes of data into the Hadoop distributed file system (HDFS).

- **Text Analytics.** Text analytics is a method for extracting usable knowledge from unstructured text data through the identification of core concepts, sentiments, and trends, and then using this knowledge to support decision making. Text analytics helps in contextualizing the unstructured data.

- **Big Data Discovery.** Big data platforms enable experimentations and discovery processes. The discovery process is a key function that helps in determining patterns in the data.

- **Big Data Quality.** Data quality management is a discipline that includes the methods to measure and improve the quality and integrity of an organization's data. However, big data quality will require radically different approaches from a technology perspective. For example, organizations may need to consider the following approaches:

  1. Address data quality natively within Hadoop.

  2. Leverage unstructured content to improve the quality of sparse data.

  3. Use CEP to improve data quality in real-time without landing data to disk.

- **Big Data Metadata.** Metadata is information that describes the characteristics of any data object, such as its name, location, perceived importance, quality, or value to the enterprise, and its relationships to other data objects that the enterprise deems worth managing. Big data expands the volume, velocity, and variety of information while adding new challenges in building and maintaining a coherent metadata infrastructure. As organizations store more of their data within Hadoop, they will need to address data lineage and impact analysis within this environment as well.

- **Master Data Management.** Organizations may want to enrich their master data with additional insight from big data. For example, they might want to link social media sentiment analysis with master data to understand if a certain customer demographic is more favorably disposed to the company's products. Organizations will also need well-governed, clean reference data such as codes for gender, countries, states, currencies, and diseases, to support their big data projects. All the major MDM vendors also offer tools to manage reference data.

- **Information Policy Management.** Information governance is all about managing information policies. Whether they recognize it or not, organizations grapple with five important processes relating to information policies:

  1. Documenting policies relating to data quality, metadata, privacy, and information lifecycle management. For example, a big data policy might state that call center agents should not record social security numbers in their notes.

  2. Assigning roles and responsibilities such as data stewards, data sponsors, and data custodians.

3. Monitoring compliance with the data policy. In the abovementioned example, the organization might use text analytics tools to identify instances where call center agents' notes contain social security numbers.

4. Defining acceptable thresholds for data issues. In the example, the information governance team might determine that the acceptable threshold needs to be zero instances because of the potential privacy implications of having social security numbers in clear text.

5. Managing issues especially those that are long-lived and affect multiple functions and lines of business. Taking the example further, the information governance team might create a number of trouble tickets so that the customer service team can eliminate any mentions of social security numbers within agents' notes.

- **Big Data Warehouses and Enterprise Data Warehouse.** As organizations adopt big data, they will increasingly follow a hybrid approach to integrate Hadoop and other NoSQL technologies with their traditional data warehousing environments.

- **Big Data Analytics.** Analytics models will increasingly incorporate big data types. Besides development of sophisticated algorithms for structured data that can work on large volumes of data, you will need analytics capabilities for the unstructured data types as well. Concepts like social listening specialized analytics on streaming data are critical for the big data platforms.

- **Big Data Reporting and Advanced Data Visualization.** Traditional reporting solutions will not work on the scale and variety of data types as big data. You will need advanced data visualization solutions to visualize and analyze big data.

- **Big Data Lifecycle Management.** Information lifecycle management (ILM) is a process and methodology for managing information through its lifecycle, from creation through disposal, including compliance with legal, regulatory, and privacy requirements. The components of a big data lifecycle management platform are listed below:

1. *Information archiving.* As big data volumes grow, organizations need solutions that enable efficient and timely archiving of structured and unstructured information while enabling its discovery for legal requirements, and its timely disposition when no longer needed by the business, legal, or records stakeholders.

2. *Records and retention management.* Every ILM program must maintain a catalog of laws and regulations that apply to information in the jurisdictions in which a business operates. These laws, regulations, and business needs drive the need for a retention schedule that determines how long documents should be kept and when they should be destroyed. Records management solutions enforce a business process around document retention.

3. *Legal Holds and Evidence Collection (eDiscovery).* Most corporations and entities are subject to litigation and governmental investigations that require them to preserve potential evidence. Large entities may have hundreds or thousands of open legal matters with varying obligations for data. Data sources include e-mail, instant messages, Excel spreadsheets, PDF documents, audio, video, and social media.

4. *Test Data Management.* The big data governance program needs tools to streamline the creation and management of test environments, subset and migrate data to build realistic and right-sized test databases, mask sensitive data, automate test result comparisons, and eliminate the expense and effort of maintaining multiple database clones.

- **Big Data Security and Privacy.** Since big data platforms provide a wide array of possibilities to access any data types (internal and external), the questions around ethical usage of data, data security, and privacy become critical. A big data platform should make positions to ensure security and privacy of data following some of the methods outlined below:

    1. *Data Masking.* These tools are critical to de-identify sensitive information, such as birth dates, bank account numbers, street addresses, and Social Security numbers.

    2. *Database Monitoring.* These tools enforce separation of duties and monitor access to sensitive big data by privileged users. The database monitoring functionality must have a minimal impact on database performance and should not require any changes to databases or applications.

- **Cloud.** Organizations are also turning to the cloud because of perceived flexibility, faster time-to-deployment, and reduced capital expenditure requirements. A number of vendors offer big data platforms in the cloud and we list a few examples below:

End Points

The big question for enterprises with growing big data analytics investments is whether to choose an RDBMS-only solution, or a dual RDBMS and map-reduce/Hadoop solution. Over time, the two architectures will not exist as separate islands but rather will have rich data pipelines going in both directions. It is safe to say that both architectures will evolve hugely over the next decade.

Sometimes when an exciting new technology arrives, there is a tendency to close the door on older technologies as if they were going to go away. Traditional data warehousing has built an enormous legacy of experience, best practices, supporting structures, technical expertise, and credibility with the business world. This will be the foundation for information management in the upcoming decade as data warehousing expands to include big data analytics.

A next-generation data architecture is emerging that connects the classic systems powering business transactions and interactions with Hadoop, a hybrid architecture capable of storing, aggregating, and transforming multi-structured raw data sources into usable formats that help fuel new insights for the business. The unprecedented growth and availability of data across a diverse set of channels and the competitive advantage that organizations gain from harnessing that data are the key driving factors for big data adoption. Hadoop's ability to run on commodity servers, store a broad range of data types, process analytic queries via map-reduce and predictably scale with increased data volumes are very attractive solution characteristics as it pertains to big data analytics. RDBMS based EDW solutions such as *Netezza* and *Greenplum* appliances enable low latency access to high volumes of data, provide data retrieval via SQL, integrate with a wide variety of enterprise BI and ETL tools and are optimized for price/performance across a diverse set of workloads. Organizations that architect their big data platforms integrating the two technologies have the ability to take advantage of the best of both worlds.

Big data analytics is a computational discipline and one would need to skillfully architect multiple technologies to meet its broad objectives. It's disruptive in nature and would pose architectural challenges to IT organizations similar in scale as SOA in the late 1990s and cloud computing over the last decade. Organizations that overcome those challenges and use the right set of technologies for big data analytics will be successful.

References

Big Data: Hadoop, Business Analytics and Beyond: Jeff Kelly Nov 08, 2012:
  http://wikibon.org/wiki/v/Big_Data:_Hadoop,_Business_Analytics_and_Beyond
CAP Twelve Years Later: How the "Rules" Have Changed: Eric Brewer on May 30, 2012 -
  http://www.infoq.com/articles/cap-twelve-years-later-how-the-rules-have-changed
Key Value Database: bigdatanerd.wordpress.com
NoSQL Databases: www.newtech.about.com
DataWarehouseBigDataAnalyticsKimball.pdf:
  http://www.montage.co.nz/assets/Brochures/
  DataWarehouseBigDataAnalyticsKimball.pdf
Big Data Diversity Meets EDW Consistency for New Synergies in BI: Nancy McQuillen,
  2 December 2011

Big Data: Data Quality's Best Friend?:
http://ibmdatamag.com/2012/08/big-data-data-qualitys-best-friend
Does Big Data Need Bigger Data Quality and Data Management?: Source: http://www.verisk.
com/Verisk-Review/Articles/Does-Big-Data-Need-Bigger-Data-Quality-and-
Data-Management.html
Integrating Master Data Management with Big Data: http://www.tmcnet.com/topics/
articles/2012/07/10/298205-integrating-master-data-management-with-big-data.htm
How MDM Fits with Big Data, Mobile & Cloud: http://www.masteringdatamanagement.
com/index.php/2012/08/14/how-mdm-fits-with-big-data-mobile-cloud
Mastering the Big Data Explosion with MDM:
http://www.dataversity.net/mastering-the-big-data-explosion-with-mdm/
Master Data Management Grows Up, the Finale: Big Data: http://www.itbusinessedge.com/
blogs/integration/master-data-management-grows-up-the-finale-big-data.html
Big Data Quality: Persistence vs. Disposability: http://www.information-management.com/
blogs/big-data-quality-persistence-versus-disposable-10023136-1.html
MDM in a Big Data World:
http://www.information-management.com/blogs/mdm-in-a-big-data-
world-10023134-1.html
Master Data Management – A Foundation for Big Data Analysis:
https://blogs.oracle.com/mdm/entry/master_data_management_a_foundation
What is the Big Deal about MDM + Big Data?:
http://corrigandavid.wordpress.com/2012/04/25/what-is-the-big-deal-about-
mdm-big-data/
The emerging relationship between MDM and big data: http://corrigandavid.wordpress.
com/2012/04/03/the-emerging-relationship-between-mdm-and-big-data/
Is MDM Ready for Big Data:
http://corrigandavid.wordpress.com/2012/04/10/is-mdm-ready-for-big-data/

⬛ ⬛ ⬛

CHAPTER 6

Data Modeling Approaches for Big Data and Analytics Solutions

One common theme you will hear again and again concerning big data solutions: there is no schema to model! Does this mean we do not need to do any data modeling activities while constructing a big data solution?

Data integration, in effect is the acquisition of data from diverse source systems (like operational applications for ERP, CRM, supply chain, where most enterprise data originates and a host of external sources of data like social networks, external third party data sources, etc.) through multiple transformations of the data to get it ready for loading into target systems (like data warehouses, customer data hubs, and product catalogs). Heterogeneity is the norm for both data sources and targets, since there are various types of applications, databases, file types, and so on. All these have different data models, so the data must be transformed in the middle of the process, and the transformations themselves vary widely. Then there are the interfaces that connect these pieces, which are equally diverse. And the data doesn't flow uninterrupted or in a straight line, so you need data staging areas. Simply put, that's a lot of complex and diverse activities that you must perform to organize data to make it useful.

Eventually the data integration processes and approaches influence the data model development as well. Let's first understand the data integration patterns.

Understanding Data Integration Patterns

Data integration approaches can become highly complex especially when you are dealing with big data types. Below is an attempt to outline the complexities of data integration processes.

- **Level 0:** Simple point to point data integration with little or no transformation. This just means information is flowing from one system to another.

- **Level 1:** Simple data integration processes, transforming one schema to another, without applying any data manipulation functions like "if," "then," "else," etc.

155

- **Level 2:** Simple data integration processes, transforming one schema to another, with application of data manipulation functions like "if," "then," "else," etc.

- **Level 3:** Complex data integration patterns, transforming the subject data dealing with complex schemas and semantic management involving both structured and unstructured data. In this scenario there could be one or more data sources (data could be also at rest or in motion) and one or more schema targets.

These design patterns (and there could be many more depending on the applications you are trying to develop and the nature of data sources) need to be aligned with the right integration architectures and influence the resulting data model to a great extent. We purposefully stayed away from discussing the granularity of data, state of data changes, and governance processes around data: if you add those aspects to the data integration patterns you can realize the complexity of the solution.

Big Data Workload Design Approaches

Big data use cases range from data ingestion to complex and real-time analytics. Each one of these use cases applies a specific data processing technique and a data analysis technique (Figure 6-1). In the center lies the data model (or lack of data model). Hence it is important for us to understand these workload design patterns first before we deep dive into data modeling techniques for big data scenarios. Once we are able to categorize where the big data workloads fall with respect to a business use case it becomes easier to map the right architectural constructs required to implement the workload – *columnar, Hadoop, name value, graph databases, complex event processing (CEP) and machine learning processes.*

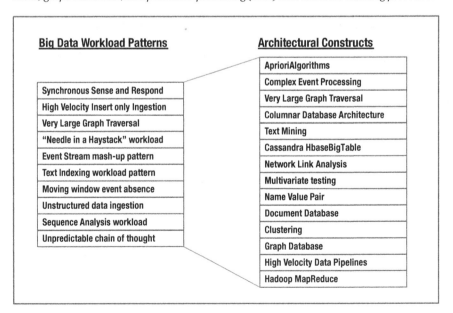

Figure 6-1. Big data workload patterns

156

- Data Workload-1: **Streaming Analytics.** This type of workload essentially consists of processing streaming data with predefined behavioral patterns at real time. Once a pattern is observed then real time responsesare formulated.

- Data Workload-2: **High Velocity Data Ingestion.** There are several subpatterns in this type of workload.

  - You can simply keep collecting the data without applying any transformations; this data at a later point in time can be analyzed. The intent is to not to lose the data streams as they happen.

  - In other scenarios, you may have to collect the data, transform and analyze all at the same time to contextualize the data.

- Data Workload-3: **Linkage Analysis.** Primarily these types of workloads are meant to establish relationships and linkages between different states of data. This workload is computation and read intensive as node statistics need to be computed and children of a node need to be read dynamically.

- Data Workload 4: **Rare-Event Detection.** Looking for a specific pattern from the vast data sets across multiple attributes is a very dataanalysis workload.

- Data Workload 5: **Data Mash-Ups.** Usually in these types of workloads you are developing a story line or creating a "data bag" linking not only data attributes but also events that happened in isolation and may not have significance. But taken together as a string of events occurring in a timeline, their importance amplifies especially across multiple event streams. Sequence analysis linking pieces of events together are some of the common examples of these types of workloads.

- Data Workload 6: **Text Analytics.** A very commonly observed data workload in big data scenarios: sentiment analysis, opinion mining, social network analysis etc., fall majorly in this category.

- Data Workload-7: **Time Series Analysis.** This type of work loads deals with pattern detections and occurrence or non-occurrence of specific events across moving time windows of data.

- Data Workload-8: **Data Forensic.** This workload is primarily triggered by data scientists exploring large data sets with questions previously not thought of. They cast a wide net and often come up with few patterns. The query patterns in this type of data workload are both "depth search" as well as "breadth search."

Map-Reduce Patterns, Algorithms, and Use Cases

In this section we will discuss a number of map-reduce patterns and algorithms to give a systematic view of the different techniques that can be found on the Web or scientific articles. Several practical examples are also provided.

Map-Reduce Patterns by Example

Let's take the example of customer and orders and draw an analogy to the map-reduce way of doing queries to address a few business specific questions.

Figure 6-2 illustrates the customer-order relationship. In an e-commerce or supermarket scenario you will encounter millions of orders, hence as a database design consideration we have sharded the order data set over a large cluster of nodes. Our sales analysis folks want to see a product and its total revenue for the last seven days. In this case, to prepare the revenue at a product level, if we follow the traditional SQL approach, we will have to query across many nodes. However, Let's explore how to apply the map-reduce pattern to solve this case.

Customer ID	Customer Name	Shipping Address	Billing Address
1	Soumendra	...	...

Order ID	Customer ID	Line Item Name	Unit Price	Quantity	Total Price
1	1	Nike Shoes	400$	1	400$
1	1	Shaving Cream	12$	1	12$
1	1	Men's Perfume -X	28$	2	56$

Figure 6-2. Customer: Order illustration

The first stage in a map-reduce job is the map stage. A map is a function whose input is a single aggregate (read as group by) and whose output is a bunch of key-value pairs. Taking our example into account, the input is an item and the output is the key-value pairs (total price and quantity) corresponding to the line items.

Figure 6-3 illustrates our map function.

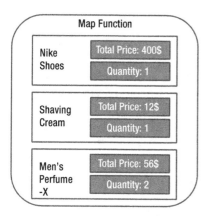

Figure 6-3. *Map function applied to customer: Order illustration*

In order to provide high parallelization, map functions operate on a single record independent of all others. The reduce function (Figure 6-4) takes multiple map outputs with the same key and combines their values to arrive at the final result.

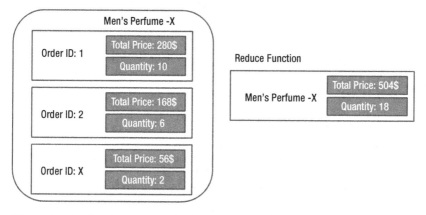

Figure 6-4. *Reduce function applied to customer: Order illustration*

The map-reduce framework arranges for map tasks to be run on the correct nodes to process all the data sets and for the data to be moved to the reduce function. In its simplest form, you can think of the map-reduce job having a single reduce function, the outputs from all the map tasks running on various nodes are then aggregated together and sent to the reduce function.

What optimization options do we have for the map-reduce framework? Each reduce function operates on the results of a single key, so this limits the performance; you can't do anything in the reduce function to make it operate across keys. On the other hand this limitation is actually a good thing; it allows you to run multiple reducers in parallel.

If we have to improve on parallelism, we will have to divide the results of the mapper based on the key on each processing node (typically multiple keys are grouped together into partitions), then we take data from all the nodes for one partition, combine it into a single group for that partition and send it to the reducer. Multiple reducers can then operate on the partitions in parallel to get to the final results merged together (Figure 6-5). This approach sometimes is called "shuffling" and the partitions are referred to as "buckets" or "regions."

Parallel Reduce Functions by Partitions

Map Function Output (By Order)

Men's Perfume-X	280$
Men's Perfume –X	168$
Shaving Cream	12$
Shaving Cream	48$
Men's Perfume -X	56$
Nike Shoes	400$
Nike Shoes	286$
Men's Perfume –X	28$
Shaving Cream	24$

Men's Perfume - X	280$
Men's Perfume –X	168$
Men's Perfume -X	56$
Men's Perfume -X	28$

Shaving Cream	12$
Shaving Cream	48$
Shaving Cream	24$

Nike Shoes	400$
Nike Shoes	286$

Figure 6-5. Parallel reduce functions applied to customer: Order illustration

While we optimized the map-reduce function through parallel reduce functions, we still see data being moved from node to node between the map and reduce functions. How can we optimize the process to minimize data movements?

If you notice, much of this data is repetitive, consisting of multiple key-value pairs for the same key. We can introduce a combiner function to the map-reduce framework, which will combine all the data for the same key into a single value.

■ **Note** A combiner function in essence is a reducer function.

Map-reduce framework imposes few limitations on the calculations you would like to perform on the data. Within a map function you can only operate on a single aggregate, and within a reduce function, you can only operate on a single key (Figure 6-6). Thus, based on your requirement, you will have to design different map-reduce jobs. To illustrate this aspect, let's look at the requirement: "What is the average ordered quantity for each product?"

Map Function Output (By Order)

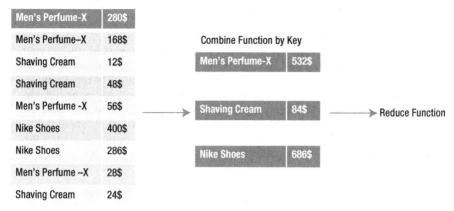

Figure 6-6. *Combiner function applied to customer: Order illustration*

An important property of averages is that they are not additive, meaning we can't take two groups of orders, find the average quantity for products in these orders and then add all the averages for the products to get to our answer. The right way of doing this is to take total quantity for a product from each order, add these figures up and then calculate the average from the combined sum (Figure 6-7).

Combine Function by Product – Node 1

Men's Perfume-X	
Total Quantity	16
Total Orders	4
Mean Quantity	4.0

Combine Function by Product – Node 2

Men's Perfume-X	
Total Quantity	42
Total Orders	12
Mean Quantity	2.5

Reduce Function by Product – All Nodes

Men's Perfume-X	
Total Quantity	94
Total Orders	25
Mean Quantity	3.07

Combine Function by Product – Node X

Men's Perfume-X	
Total Quantity	36
Total Orders	9
Mean Quantity	4.0

Figure 6-7. *Average orderd quantity for a product applied to customer: Order illustration*

161

The examples we have discussed above are complete map-reduce computations, where we start from raw input data and create a final output. Many map-reduce functions take a while to perform, even with clustered nodes. As new data keeps coming in, you will have to re-run the map-reduce computations to stay up to date.

The map stages of the map-reduce are easy to handle incrementally; you run your mapper function only if the input data has changed, since map functions are isolated from each other, handling incremental updates are straightforward. The more complex case is the reduce step, since it pulls together all the outputs from many maps and any changes in the map outputs necessitates a re-run of the reduce function. This issue can be resolved depending upon how parallel the reduce step is. If we are partitioning data for reduction, then any partition that remains unchanged and does not necessitate the reduce function to re-run on that partition.

Basic Map-Reduce Patterns
Counting and Summing

Problem Statement: There are a number of documents where each document is a set of terms. It is required to calculate a total number of occurrences of each term in all documents. Alternatively, it can be an arbitrary function of the terms. For instance, there is a log file where each record contains a response time and it is required to calculate an average response time.

Applications:

Log Analysis, Data Querying

Collating

Problem Statement: There is a set of items and some function of one item. It is required to save all items that have the same value of function into one file or perform some other computation that requires all such items to be processed as a group. The most typical example is building of inverted indexes.

Solution: The solution is straightforward. Mapper computes a given function for each item and emits value of the function as a key and item itself as a value. Reducer obtains all items grouped by function value and process or save them. In case of inverted indexes, items are terms (words) and function is a document ID where the term was found.

Applications:

Inverted Indexes, ETL

Filtering ("Grepping"), Parsing, and Validation

Problem Statement: There is a set of records, and it is required to collect all records that meet some condition or transform each record (independently from other records) into another representation. The latter case includes such tasks as text parsing and value extraction, conversion from one format to another.

Solution: Solution is absolutely straightforward: mapper takes records one by one and emits accepted items or their transformed versions.
Applications:

> Log Analysis, Data Querying, ETL, Data Validation

Distributed Task Execution

Problem Statement: There is a large computational problem that can be divided into multiple parts and results from all parts can be combined together to obtain a final result.

Solution: Problem description is split into a set of specifications, and specifications are stored as input data for mappers. Each mapper takes a specification, performs corresponding computations, and emits results. Reducer combines all emitted parts into the final result.
Applications:

> Physical and Engineering Simulations, Numerical Analysis, Performance Testing

Sorting

Problem Statement: There is a set of records, and it is required to sort these records by some rule or process these records in a certain order.

Solution: Simple sorting is absolutely straightforward: mappers just emit all items as values associated with the sorting keys that are assembled as function of items. Nevertheless, in practice sorting is often used in a tricky way, that's why it is said to be the heart of map-reduce (and Hadoop). In particular, it is very common to use composite keys to achieve secondary sorting and grouping. Sorting in map-reduce is originally intended for sorting of the emitted key-value pairs by key, but there exist techniques that leverage Hadoop implementation specifics to achieve sorting by values.

It is worth noting that if map-reduce is used for sorting of the original (not intermediate) data, it is often a good idea to continuously maintain data in sorted state using BigTable concepts. In other words, it can be more efficient to sort data once during insertion than sort them for each map-reduce query.
Applications:

> ETL, Data Analysis

Advanced Map-Reduce Patterns
Iterative Message Passing (Graph Processing)

Problem Statement: There is a network of entities and relationships between them. It is required to calculate a state of each entity on the basis of properties of the other entities in its neighborhood. This state can represent a distance to other nodes, indication that there is a neighbor with the certain properties, characteristic of neighborhood density and so on.

Solution: A network is stored as a set of nodes, and each node contains a list of adjacent node IDs. Conceptually, map-reduce jobs are performed in iterative way, and at each iteration each node sends messages to its neighbors. Each neighbor updates its state on the basis of the received messages. Iterations are terminated by some condition like fixed maximal number of iterations (say, network diameter) or negligible changes in states between two consecutive iterations. From the technical point of view, mapper emits messages for each node using ID of the adjacent node as a key. As result, all messages are grouped by the incoming node and reducer is able to re-compute state and rewrite node with the new state.

Example: In an ecommerce application, there is a tree of categories that branches out from large categories (like men, women, kids) to smaller ones (like men's jeans or women's dresses), and eventually to small end-of-line categories (like men's blue jeans). End-of-line category is either available (contains products) or not. Some high level category is available if there is at least one available end-of-line category in its sub tree. The goal is to calculate availabilities for all categories if availabilities of end-of-line categories are known.

Example: PageRank and mapper-side data aggregation, this algorithm was suggested by Google to calculate relevance of a web page as a function of authoritativeness (PageRank) of pages that have links to this page. The real algorithm is quite complex, but in its core it is just a propagation of weights between nodes where each node calculates its weight as a mean of the incoming weights.

Applications:

Graph Analysis, Web Indexing

Distinct Values (Unique Items Counting)

Problem Statement: There is a set of records that contain fields F and G. Count the total number of unique values of field F for each subset of records that have the same G (grouped by G). The problem can be a little bit generalized and formulated in terms of faceted search. There is a set of records. Each record has field F and arbitrary number of category labels G = {G1, G2, ...}. Count the total number of unique values of filed F for each subset of records for each value of any label.

Applications:

Log Analysis, Unique Users Counting

Cross-Correlation

Problem Statement: There is a set of tuples of items. For each possible pair of items calculate a number of tuples where these items co-occur. If the total number of items is N then N*N values should be reported.

This problem appears in text analysis (say, items are words and tuples are sentences), market analysis (customers who buy *this* tend to also buy *that*). If N*N is quite small and such a matrix can fit in the memory of a single machine, then implementation is straightforward.

The first approach is to emit all pairs and dummy counters from mappers and sum these counters on reducer. The shortcomings are:

- The benefit from combiners is limited, as it is likely that all pair are distinct

- There is no in-memory accumulations

The second approach is to group data by the first item in pair and maintain an associative array ("stripe") where counters for all adjacent items are accumulated. Reducer receives all stripes for leading item i, merges them, and emits the same result as in the pairs approach.

- Generates fewer intermediate keys. Hence the framework has less sorting to do.

- Greatly benefits from combiners.

- Performs in-memory accumulation. This can lead to problems, if not properly implemented.

- More complex implementation.

- In general, "stripes" is faster than "pairs"

Applications:

Text Analysis, Market Analysis

NoSQL Data Modeling Techniques

SQL and the relational model in general were designed to store and manage data originating from enterprise systems and the main focus was to stay ACID compliant. While SQL and relational models ensured data integrity and consistency, they also introduced abstractions modeling end user interactions. This user-oriented nature had a few implications:

- Mostly the end user wanted to see data at aggregated level for reporting and analysis purpose. Contextualizing the data and linkages were not possible through standard SQL functions, hence complex applications needed to be built to bring out the semantic meaning of data.

- Distributed nature of data management operations was never thought of, while SQL and RDBMS platforms provided excellent features to manage concurrency, integrity, consistency, or data type validity, they fail in providing consistency, availability and fault-tolerance type of features, which was largely left to the programmer community to custom develop.

To overcome these shortcomings and most importantly to develop solutions to manage big data scale and variety of data types a new set of "No SQL" (read as Not Only SQL) data models began to emerge: *key-value storage, document databases,* and *graph databases.*

Types of NoSQL Data Stores

The following section describes the different types of NoSQL datastores.

Key-Value stores

Examples: Tokyo Cabinet/Tyrant, Redis, Voldemort, Oracle BDB
Typical applications: Content caching
Strengths: Fast lookups
Weaknesses: Stored data has no schema

Example application: The web application is an internal crowd-sourcing portal of a company, where people share innovative ideas, and there are messages posted about these ideas. In order to promote innovative thinking it is important to capture how many ideas are getting posted, how many people are commenting to those ideas, etc. The web page reads from a key that is based on the user's ID and retrieves a string of JSON that represents all the relevant information. A background process recalculates the information every 15 minutes and writes to the store, independently awarding points against the maximum polled idea.

Document databases

Examples: CouchDB, Mongobd
Typical applications: Web applications
Strengths: Tolerant of incomplete data
Weaknesses: Query performance, no standard query syntax

Example application: A collaboration interface that takes into account several people providing inputs across various phases of a design document. The details you need to capture for each activity vary tremendously with design considerations and when new requirements are shared by clients. The entire document is built up piecemeal, with each and every input needing to be captured. Multiple people are collaborating who may or may not have the big picture in front of them, so the completeness and accuracy of the design can only be ascertained when all the sections of the document are complete. But until then you have to treat the individual contributions skeptically.

Graph databases

Examples: Neo4J, InfoGrid, Infinite Graph
Typical applications: Social networking, recommendations
Strengths: Graph algorithms (e.g., shortest path, connectedness, n degree relationships, etc.)
Weaknesses: Has to traverse the entire graph to achieve a definitive answer. Not easy to cluster.

Example application: Any application that requires linkage analysis and relationship analysis across a large cluster of people is best suited to a graph database.

XML Databases

Examples: Exist, Oracle, MarkLogic
Typical applications: Publishing
Strengths: Mature search technologies, Schema validation
Weaknesses: No real binary solution, easier to re-write documents than update them

Example application: A logistics company that uses bespoke XML formats to manage the invoices, customer order fulfillment directives, shipping commitments associated with the orders and packaging and handling instructions for each order. The company manager needs to quickly search either text or semantic sections of the markup (e.g., orders whose summary contains fragile, where the shipping commitment is four business days and customers belonging to a particular zip code area). This application extensively uses text-based search techniques.

Distributed Peer Stores

Examples: Cassandra, Hbase, Riak
Typical applications: Distributed file systems
Strengths: Fast lookups, good distributed storage of data
Weaknesses: Very low-level API

Example application: A plant maintenance application that takes into account the plant machineries generating logs, where each machine's log needs to be captured separately. All these log files are processed in a batch mode every four hours to identify any specific anomalies in machine's readings, and a list of alerts gets generated for a machine that has higher-than-normal thresholds. The plant's machines are in remote places and spread a wide area.

What Database System Should Your Application Use?

The key point is to determine what the objective of your application is. Table 6-1 shows some of the basic guidelines in determining what use cases need what kind of database solutions.

Table 6-1. *Use cases for No SQL databases*

Use Case Scenario	Use Case Requirements	Recommended Database
Complex transactions	Can't afford to lose data Example – Inventory management system that needs to be fully ACID compliant	Relational or Grid database
Highly scalable	Scale-out partitioning, live addition and removal of nodes, load balancing, automatic sharding and rebalancing and fault tolerance	No SQL and RDBMS
Always be able to write	High availability and eventual consistency	Bigtable clones
Small and continuous reads and writes	Data may be volatile but needs fast streaming type of processing and in-memory accessing	Document or Key-value type of No SQL databases
Social network operations	Discover networks, relationships, pattern matching and correlations	Graph databases, Riak, Redis
Wide variety of access patterns across many different data types	Breadth search and depth search, patterns findings, long running queries, deep analytic type of usage	Document databases
Offline reporting with large data sets	Mostly for data exploratory analysis purpose	Hadoop/MapReduce
Distributed data management	Spanning multiple data centers to handle latency and are partition tolerant	Bigtable Clones
CRUD applications	Access complex data without joins	Document databases
Search applications	Content Analytics	Riak
Operations on multiple data structures	Lists, Sets, Queues, Publish-Subscribe	Redis
Programmer friendliness	Rapid Application development and deployment	Document databases and Key-value databases
Transactions oriented with real-time processing	Data roll-ups, time windowing, materialized views, real time data feeds	Volt DB

(continued)

Table 6-1. (*continued*)

Use Case Scenario	Use Case Requirements	Recommended Database
Log processing	Continuous streams of data that may have no consistency	Bigtable clones
Dynamic relationship building	Dynamically build relationships between object that have dynamic properties; will not require a rigid schema and models needs to be developed programmatically	Graph databases
Large Media – BLOB types	Support large media data types, Caching of web pages or to save complex objects that were expensive to join in a relational database, etc.	S3, Mongo DB
Bulk upload	Bulk upload lots of data quickly and efficiently	Mostly RDBMS solutions
Easier upgrade options	Fluid schema system that supports optional fields, adding fields, deleting fields without requiring to build an entire schema migration framework	Document databases or Key-value databases
Mobile platform	For multichannel data ingestion as well as data consumption	Couch DB, Mobile Couchbase

With the above information, let's look at a simplistic product table consisting of fields Product ID, Product Name, Product Categories and Product Packaging and then apply data modeling techniques as suitable to both relational databases and non-relational databases.

Let's examine the relational database model first in Figure 6-8.

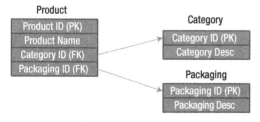

Figure 6-8. *Product RDBMS logical data model*

In contrast, the NoSQL databases take a completely different approach and implement interaction-based data structure solutions. In general, NoSQL data models can be classified into four categories as described below:

- Column-based store

- Document store

- Key-value store

- Graph data store

Column Families or Wide-column Store

Examples: Cassandra, Hbase
Typical usages: Distributed data storage.
The column-based stores extend the typical Key-Value pair storage where each column can be grouped with a key and corresponding set of values. These types of data structures are preferred when the application requires extensive read/write operations.
By following this approach our product data model will look like Figure 6-9.

Product ID	Product Name	Category		Packaging	
		Category ID	Category Desc	Packaging ID	Packaging Desc
1	Kelogg Cereal	NT 1	Cereal	BX 1	Box-24inches-Rectangular

Figure 6-9. Product column family logical data model

Later in this chapter we will discuss how to develop data models using Cassandra's column family data store.

Document Store

Examples: CouchDB, MongoDB
Typical usages: Web applications.
Mostly applicable when the data is in document format; highly unstructured in nature, these documents are stored and retrieved following the XML, JSON, or BSON architecture principles.

By following this approach our product data model will look like below:

```
Product Id = "001",ProductName="Kelogg Cereal",
Category=[{Category Id:"NT1",CategoryDesc:"Cereal"}],
Packaging=[{Packaging ID="BX1", Packaging Desc:"Box-24 inches-Rectangular"}]
```

Key-Value Store

Examples: Membase, Redis

Typical usages: Distributed hash table, caching

Key-Value store behaves like hash tables where the values are mapped to keys. The simplicity of this model allows storage and processing of any type of information, thus creating a schema-less storage. Whenever the business application needs excessive read operations or search type of interactions with data, this model is recommended.

Graph Databases

Examples: Neo4J, InfoGrid

Typical usages: Social networking, recommendations

The Graph Data model is based on Graph theory where data is stored in nodes, and the linkages to other data are reflected through the edges. This data model is recommended when the business application does a lot of recursive analysis.

By following this approach our product data model will look like below:

Node: Product

Property: Product ID, Product name, category, and packaging

Relationship: Each product is mapped to a category, and each product has a packaging specification.

▓ **Note** Relational data modeling is based on the design-themed question, "What answers do I have from the available data?" This means you must develop applications and formulate queries based on the data structures and available data. Whereas, No SQL data modeling is based on the design-themed question: "What questions do I have?" This means, irrespective of data structures, you design applications with specific questions in mind.

What is JSON

JSON (stands for JavaScript Object Notation) is a lightweight and highly portable data-interchange format. JSON is intuitive to the Web as well as the browser. Interoperability with any/all platforms in the current market can be easily achieved using JSON message format.

According to JSON.org (www.json.org), JSON is built on two structures:

- A collection of name/value pairs. In various languages, this is realized as an object, record, dictionary, structure, keyed list, hash table, or associative array.

- An ordered list of values. In most languages, this is realized as an array, list, vector, or sequence.

A typical JSON syntax is as follows:

- Data is represented in the form of name-value pairs. A name-value pair comprises a "member name" in double quotes, followed by colon ":" and the value in double quotes

- Each data member (name-value pair) is separated by comma

- Objects are held within curly ("{ }") brackets.

- Arrays are held within square ("[ ]") brackets.

JSON is significantly like XML:

- JSON is plain text data format

- JSON is human readable and self-describing

- JSON is categorized (contains values within values)

- JSON can be parsed by scripting languages like Java script

- JSON data is supported and transported using AJAX

Though JSON and XML are both data formats, JSON has few advantages over XML because of the following reasons:

- JSON is lighter compared to XML (No unnecessary/additional tags in JSON)

- JSON is easier to read and understand by humans.

- JSON is easier to parse and generate for machines.

- For AJAX related applications, JSON is faster than XML.

Column Family Database: Columns, Column Family, Super Column Family

Column family databases are probably most known because of Google's BigTable implementation. They are very similar to relational database, but they also have differences in their approach to storing and accessing data. Some of the difference is storing data by rows (relational) versus storing data by columns (column family databases). But a lot of the difference is conceptual in nature. You can't apply the same sort of solutions that you used in a relational form to a column database.

The following concepts are critical to understand how column databases work:

- Column family

- Super columns

- Column

You need to define the schema for tables in relational databases; however, the only thing that you define in a column family is the name and the key sort options (there is no schema).

- Column families. A column family is how the data is stored on the disk. All the data in a single column family will sit in the same file (actually, set of files, but that is close enough). A column family can contain super columns or columns.

- A super column is a dictionary; it is a column that contains other columns (but not other super columns).

- A column is a tuple of name, value, and timestamp.

It is important to understand that schema design in a column family database (CFDB) is of great importance; if you don't build your schema right, you literally can't get the data out. CFDB usually offers one of two forms of queries, either by key or by key range. A CFDB is meant to be distributed, and the key determines where the actual physical data would be located. Data is stored based on the sort order of the column family, and you have no real way of changing the sorting (except choosing between ascending or descending). The sort order, unlike in a relational database, isn't affected by the columns values but by the column names.

In order to clarify the concepts of column families and the type of problems they help solve, let's look at an example.

Imagine you have a database that contains census data. The person table (Figure 6-10) has one row for each person who participated in and would probably be keyed by a unique key. All singleton attributes such as date of birth, gender, address and so forth would exist in this table. Some repeating attributes like work history wouldbe normalized out into related tables. Depending upon the size of the sample, a census may take in hundreds of millions of people, and would look something like Figure 6-10.

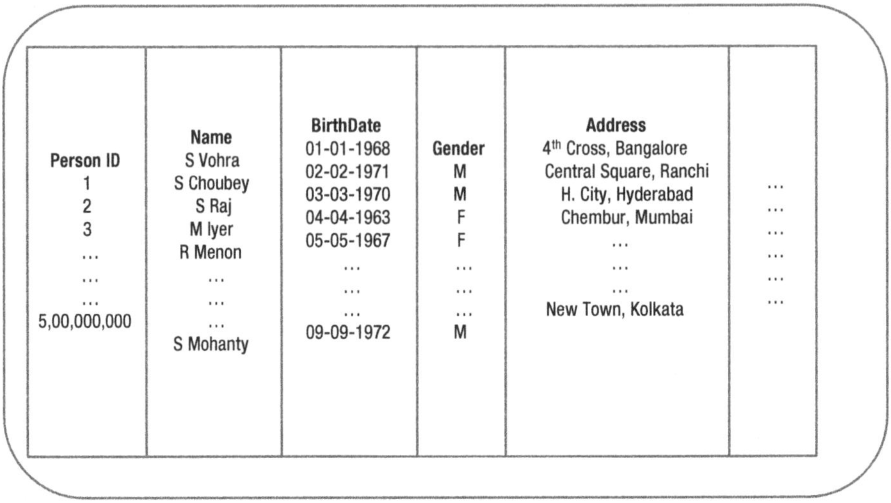

Person ID	Name	BirthDate	Gender	Address	
1	S Vohra	01-01-1968		4th Cross, Bangalore	
2	S Choubey	02-02-1971	M	Central Square, Ranchi	
3	S Raj	03-03-1970	M	H. City, Hyderabad	...
...	M Iyer	04-04-1963	F	Chembur, Mumbai	...
...	R Menon	05-05-1967	F	...	...
...	...	...	...	...	...
5,00,000,000	...	...	...	New Town, Kolkata	...
	S Mohanty	09-09-1972	M		

Figure 6-10. *Column family: Census data example*

The obvious problem is that analyzing this census data to answer a question such as "How many men were born in each year?" also entails reading the name and address of each person, together with whatever other data is present in each row.

Columnar databases were devised to solve this problem. These databases store each column separately so that aggregate operations for one column of the entire table are significantly quicker than the traditional row storage model. The problem with this columnar approach is that getting all the data for a single person becomes very expensive because the database must fetch data from numerous places on disk and glue together all those columns to represent a single row.

In contrast, in a CFDB design, columns of related data are grouped together within one table as shown in Figure 6-11. The person table has now been subdivided so that all personal name and address data is grouped together, as is statistical demographic data for each person. Any other columns in the table would be grouped accordingly as well.

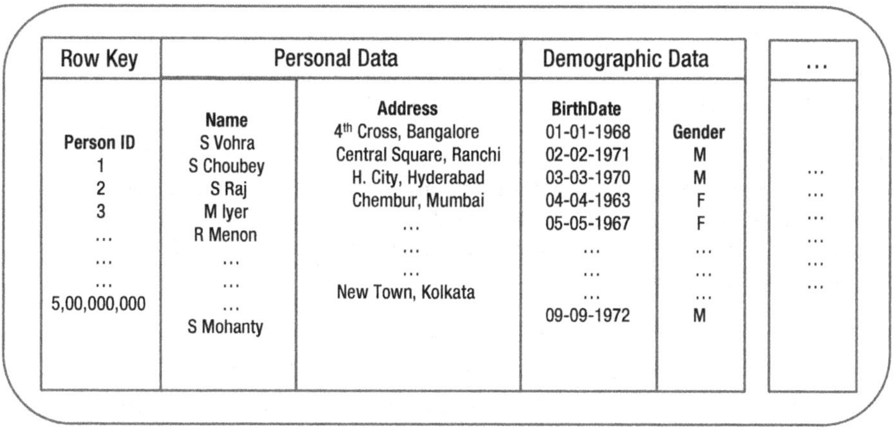

Row Key	Personal Data			Demographic Data		...
Person ID 1 2 3 5,00,000,000	**Name** S Vohra S Choubey S Raj M Iyer R Menon ... ... S Mohanty	**Address** 4th Cross, Bangalore Central Square, Ranchi H. City, Hyderabad Chembur, Mumbai ... ... New Town, Kolkata	**BirthDate** 01-01-1968 02-02-1971 03-03-1970 04-04-1963 05-05-1967 09-09-1972	**Gender** M M F F ... ... M	... ...

Figure 6-11. Column family: Census data example

NoSQL databases that support column families, such as HBase and Cassandra, only fetch the column families of those columns that are required by a query. This allows our previous statistical query to group men by age without fetching the long text columns that are present in each row. Note that the row key is not in a column family, but cuts a horizontal slice across all columns and their families to unify all data for one row.

Furthermore, NoSQL databases group all the columns in a column family together on the disk. This means they can fetch multiple rows of one column family in a single read operation. Grouping the columns like this to speed up the reading of data is called "data locality."

Now that we understand the concept of column families, let's look at a few other interesting concepts. A normal RDBMS column has a static name and a single value per row. In a CFDB design, however, a family can also be used as a container for columns where the column name itself contains data. Take a look at Figure 6-12.

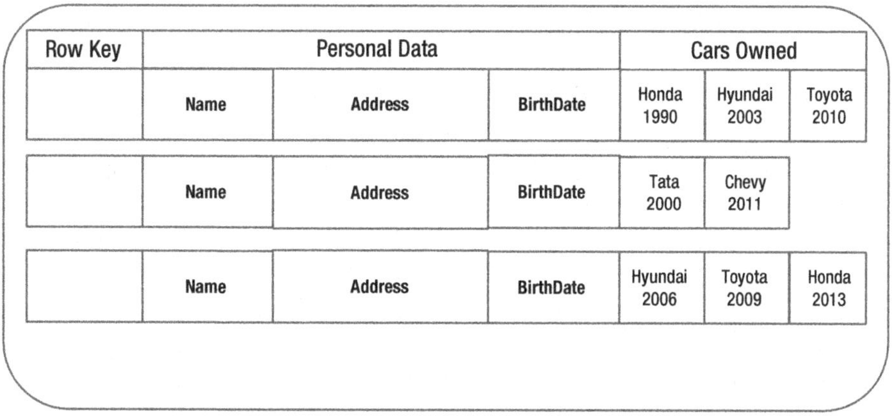

Row Key	Personal Data			Cars Owned		
	Name	**Address**	**BirthDate**	Honda 1990	Hyundai 2003	Toyota 2010
	Name	**Address**	**BirthDate**	Tata 2000	Chevy 2011	
	Name	**Address**	**BirthDate**	Hyundai 2006	Toyota 2009	Honda 2013

Figure 6-12. Column family: Data column names example

In this example, a column family in our person table is a list of cars the person has owned, and the year of manufacture for each car. This would mean a sub-table in a relational data model; but a CFDB column family can accommodate this because it can contain many name/value pairs, where the name is the column name and the value is the value of that column for that row. It is important to realize that the names of the columns in a single family can vary arbitrarily for each row.

The column families thus can be divided into static and dynamic families. Static families like personal data and demographic data in our examples above have mostly the same column names on every row. Dynamic families like cars owned contain mostly different column names for each row.

■ **Note**   A CFDB design seems to have few design considerations that are fundamental to data access: no joins, no real querying capability (except by primary key), nothing like the richness that we get from a relational database. Why is it so limited?

A CFDB is designed to run on a large number of machines and to store a *huge* amount of information. You literally cannot store that amount of data in a relational database, and even multi-machine relational databases, such as Oracle RAC, will struggle to handle the size of data and queries that are typical for CFDB.

The reason that a CFDB design doesn't provide joins is that joins require you to be able to scan the entire data set. That requires either someplace that has a view of the whole database (resulting in a bottleneck and a single point of failure) or actually executing a query over all machines in the cluster. Since that number can be pretty high, you would want to avoid such situations.

CFDB designs don't provide a way to query by column or value because that would necessitate either an index of the entire data set (or just in a single column family), which again is not practical, or running the query on all machines, which is not possible. By limiting queries to just those done by key, a CFDB design ensures that it knows exactly what node a query can run on. It means that each query is running on a small set of data, making them much cheaper and faster.

Model Column Families Around Query Patterns

As discussed earlier, No SQL data modeling is always based on query patterns; however, it is also important to understand the business context behind the objects of interest: hence, start your design with entities and relationships, if you can. Unlike in relational databases, it's not easy to tune or introduce new query patterns in by simply creating secondary indexes or building complex SQLs (using joins, order by, group by) because of its high-scale distributed nature. So think about query patterns up front, and design column families accordingly.

Entities and their relationships still matter (unless the use case is special, perhaps storing logs or other time series data). What if you are given query patterns to create a data model for an e-commerce website, but you were not told anything about the entities and relationships? You might try to figure out entities and relationships, knowingly or unknowingly, from the query patterns or from your prior understanding of the domain (because entities and relationships are how we perceive the real world). It's important to understand and start with entities and relationships, then continue modeling around query patterns by de-normalizing and duplicating.

It also helps to identify the most frequent query patterns and isolate the less frequent. Some queries might be executed only a few thousand times, while others will be executed a billion times. Also consider which queries are sensitive to latency and which are not. Make sure your model first satisfies the most frequent and critical queries.

De-normalize and Duplicate for Read Performance

In the relational world, the pros of normalization are well understood: less data duplication, fewer data modification anomalies, conceptually cleaner, easier to maintain, and so on. The cons are also understood: queries might perform slowly if many tables are joined, etc. The same holds true in column family databases, but the cons are magnified since it's a distributed database and of course there are no joins (since it's high-scale distributed). So with a fully normalized schema, reads may perform much worse.

Example: "Like" relationship between user and item

This example concerns the functionality of an e-commerce application where users can like one or more items. One user can like multiple items, and one item can be liked by multiple users, leading to a many-to-many relationship as shown in the relational model in Figure 6-13.

User ID	Name	Email
123	ABC	ABC@123.com
456	XYZ	XYZ@456.com

ID	User ID	Item ID	Timestamp
1	123	111	1234567890
2	123	222	1234560987
3	456	111	4567098123
4	456	333	7890654321

Item ID	Title	Desc
111	iPhone	Apple iPhone
222	Tipping Point	Book – Malcom Gladwell
333	Kindle	eReader

Figure 6-13. *Logical data model in RDBMS*

For this example, let's say we would like to query data as follows:

- Get user by user id

- Get item by item id

- Get all the items that a particular user likes

- Get all the users who like a particular item

Below are some options for modeling the data in CFDB, in order of the lowest to the highest de-normalization. The best option depends on the query patterns, as you'll soon see.

Option 1: Relational model

This model, Figure 6-14, supports querying user data by user ID and item data by item ID. But there is no easy way to query all the items that a particular user likes or all the users who like a particular item.

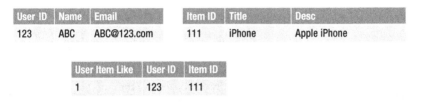

User ID	Name	Email
123	ABC	ABC@123.com

Item ID	Title	Desc
111	iPhone	Apple iPhone

User Item Like	User ID	Item ID
1	123	111

Figure 6-14. Option 1: Logical data model in RDBMS)

Option 2: Normalized entities with custom indexes

This model, in Figure 6-15, has fairly normalized entities, except that user ID and item ID mapping is stored twice, first by item ID and second by user ID.

User ID	Name	Email
123	ABC	ABC@123.com

Item ID	Title	Desc
111	iPhone	Apple iPhone

User by Item	User ID	User ID
111	123	456

Item by User	Item ID	Item ID
123	111	222

Figure 6-15. Option 2: Logical NoSQL data model

Here, we can easily query all the items that a particular user likes using "Item byUser," and all the users who like a particular item using "User byItem." We refer to these column families as custom secondary indexes, but they're just other column families.

Let's say we always want to get the item title in addition to the item ID when we query items liked by a particular user. In the current model, we first need to query "Item byUser" to get all the item IDs that a given user likes; and then for each item ID, we need to query the item to get the title. Similarly, let's say we always want to get all the usernames in addition to user IDs when we query users who like a particular item. With the current model, we first need to query "User byItem" to get the IDs for all users who like a given item; and then for each user ID, we need to query "User" to get the username. It's possible that one item is liked by a couple hundred users, or an active user has liked many items: this will cause many additional queries when we look up usernames who like a given item and vice versa. So, it's better to optimize by de-normalizing the item title in "ItembyUser" and username in "UserbyItem" as shown in option 3 (Figure 6-16).

User ID	Name	Email
123	ABC	ABC@123.com

Item ID	Title	Desc
111	iPhone	Apple iPhone

User by Item	User ID	User ID
111	123	456
	ABC	XYZ

Item by User	Item ID	Item ID
123	111	222
	iPhone	Tipping Point

Figure 6-16. *Option 3: Logical NoSQL data model*

Option 3: Normalized entities with de-normalization into custom indexes

In this model, title, and username are de-normalized in "User byItem" and "Item byUser" respectively. This allows us to efficiently query all the item titles liked by a given user and all the user names who like a given item. This is a fair amount of de-normalization for this use case.

What if you want to get all the information (title, description, price, etc.) about the items liked by a given user? First you need to ask yourself whether you really need this query, particularly for this use case. You can show all the item titles that a user likes and pull additional information only when the user asks for it (by clicking on a title). So, it's better not to do extreme de-normalization for this use case. Let's consider the following two query patterns:

- For a given "Item Id", get all of the item data (title, description, etc.) along with the names of the users who liked that item.

- For a given "User Id", get all of the user data along with the item titles liked by that user.

These are reasonable queries for item detail and user detail pages in an application. Both will perform well with this model. Both will cause two lookups, one to query item data (or user data) and another to query user names (or item titles). As the user becomes more active (starts liking thousands of items, for example) or the item becomes hotter (liked by a few million users, for example), the number of lookups will not grow; it will remain constant at two. That's not bad, and de-normalization may not yield much benefit like we had when moving from option 2 to option 3. However, you will learn how to optimize further in option 4 (Figure 6-17).

User ID	User Info		Likes	
123	Name	Email	111	222
	ABC	ABC@123.com	iPhone	Tipping Point

Item ID	Item Info		Liked By	
111	Title	Desc	123	456
	iPhone	Apple iPhone	ABC	XYZ

Figure 6-17. *Option 4: Logical NoSQL data model*

Option 4: Partially de-normalized entities

We've used the term "partially de-normalized" here because we're not de-normalizing all item data into the User entity or all user data into the Item entity.

We've left out timestamp, but let's include it in the final model, shown in Figure 6-18. Note that "time-uuid" and "userid" together form a composite column key in "User byItem" and "Item byUser" column families.

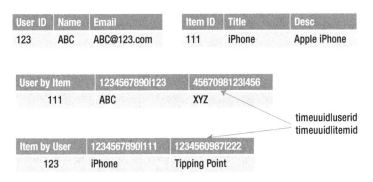

Figure 6-18. *Timestamp-enabled NoSQL data model*

Recall that column keys are physically stored sorted. Here our column keys are stored sorted by time-uuid in both "User by Item" and "Item by User", which makes range queries on time slots very efficient. With this model, we can efficiently query (via range scans) the *most recent* users who like a given item and the *most recent* items liked by a given user, without reading all the columns of a row.

We've covered a few fundamental practices and walked through a detailed example to help you get started with CFDB data model design. Here are the key takeaways:

- Don't think of a relational table, but think of a nested sorted map data structure while designing column families.

- Model column families around query patterns. But start your design with entities and relationships if you can.

- De-normalize and duplicate for read performance. But don't de-normalize if you don't need to.

- Remember that there are many ways to model. The best way depends on your use case and query patterns.

What we have not mentioned here are special-yet-common use cases such as logging, monitoring, real-time analytics (rollups, counters), or other time series data. However, the practices we discussed here do apply there. In addition, there are known common techniques or patterns used to model these time series data in a CFDB design.

Storing Values in Column Names

It's a common practice with a CFDB design to store a value (actual data) in the column name (a.k.a. column key), and even to leave the column value field empty if there is nothing else to store. One motivation for this practice is that column names are stored physically sorted, but column values are not.

Notes

- The maximum column key (and row key) size is 64KB. However, don't store something like "item description" as the column key!

- Don't use timestamp alone as a column key. You might get colliding timestamps from two or more app servers writing to CFDB. Prefer time-uuid instead.

- The maximum column value size is 2 GB. But because there is no streaming and the whole value is fetched in heap memory when requested, limit the size to only a few MBs.

Leverage Wide Rows for Ordering, Grouping, and Filtering

This goes along with the above practice. When actual data is stored in column names, we end up with wide rows.

Benefits of wide rows

- Since column names are stored physically sorted, wide rows enable ordering of data and hence efficient filtering (range scans). You'll still be able to efficiently look up an individual column within a wide row, if needed.

- If data is queried together, you can group that data up in a single wide row that can be read back efficiently, as part of a single query. As an example, for tracking or monitoring some time series data, we can group data by hour/date/machines/event types (depending on the requirements) in a single wide row, with each column containing granular data or roll-ups.

- Wide row column families are heavily used (with composite columns) to build custom indexes in CFDB.

- As a side benefit, you can de-normalize a one-to-many relationship as a wide row without data duplication.

Example

Let's say we want to store some event log data and retrieve that data hourly. As shown in Figure 6-19, the row key is the hour of the day, the column name holds the time when the event occurred, and the column value contains payload. Note that the row is wide and the events are ordered by time because column names are stored sorted. Granularity of the wide row (for this example, per hour rather than every few minutes) depends on the use case, frequency of events, and data size, etc.

ddmmyyhh	Timeuuid1	Timeuuid2
	Event Data 1	Event Data 2
	...	...

Figure 6-19. Event log logical data model

It's hard to say exactly how wide a wide row should be, partly because it's dependent upon the use case. But here's some advice:

Traffic: All of the traffic related to one row is handled by only one node/shard (by a single set of replicas, to be more precise). Rows that are too "fat" could cause hot spots in the cluster: usually when the number of rows is smaller than the size of the cluster, or when wide rows are mixed with skinny ones, or some rows become hotter than others. However, cluster load balancing ultimately depends on the row key selection; conversely, the row key also defines how wide a row will be. So load balancing is something to keep in mind during design.

Size: As a row is not split across nodes, data for a single row must fit on disk within a single node in the cluster. However, rows can be large enough so that they don't have to fit in memory entirely. Best practice is to model data in such a way that you never hit more than a few million columns or a few megabytes in one row. (In such cases where the rows are really wide, you can change the row key granularity, or you can split into multiple rows.) However, these caveats don't mean you should not use wide rows; just don't go extra wide.

Choose the Proper Row Key – It's Your "Shard Key"

Let's consider again the above example of storing time series event logs and retrieving them hourly (Figure 6-20). We picked the hour of the day as the row key to keep one hour of data together in a row. But there is an issue: All of the writes will go only to the node holding the row for the current hour, causing a hot spot in the cluster. Reducing granularity from hour to minutes won't help much, because only one node will be responsible for handling writes for whatever duration you pick. As time moves, the hot spot might also move, but it won't go away!

ddmmyyhh|EventType	Timeuuid1	Timeuuid2
	Event Data 1	Event Data 2
	...	...

Figure 6-20. Event type integrated logical NoSQL data model

One way to alleviate this problem is to add something else to the row key: an event type, machine ID, or similar value that's appropriate to your use case.

Note that now we don't have global time ordering of events, across all event types, in the column family. However, this may still be a manageable approach if the data is viewed (grouped) by event type later. If the use case also demands retrieving all of the events (irrespective of type) in time sequence, we need to do a multi-get for all event types for a given time period, and honor the time order when merging the data in the application.

If you can't add anything to the row key or if you absolutely need "time period" as a row key, another option is to shard a row into multiple (physical) rows by manually splitting row keys: "ddmmyyhh | 1", "ddmmyyhh | 2",… "ddmmyyhh | n", where n is the number of nodes in the cluster. For an hour window, each shard will now evenly handle the writes; you need to round-robin among them. But reading data for an hour will require multi-gets from all of the splits (from the multiple physical nodes) and merging them in the application.

Keep Read-Heavy Data Separate from Write-Heavy Data

Irrespective of caching and even outside the NoSQL world, it's always a good practice to keep read-heavy and write-heavy data separate because they scale differently.

Notes

- A row cache is useful for skinny rows, but harmful for wide rows today because it pulls the entire row into memory.

- Even if you have lots of data (more than available memory) in a column family but you also have particularly "hot" rows, enabling a row cache might be useful.

Make Sure Column Key and Row Key are Unique

- In many column family databases, there is no unique constraint enforcement for row key or column key.

- Also, there is no separate update operation (no in-place updates!). It's always an upsert (insert-update). If you accidentally insert data with an existing row key and column key, the previous column value will be silently overwritten without any error (the change won't be versioned; the data will be gone).

Use the Proper Comparator and Validator

In column family databases, the data type for a column *value* (or row key) is called a *validator*. The data type for a column *name* is called a *comparator*. Although the database does not require you to define both, you must at least specify the comparator unless your column family is static (that is, you're not storing actual data as part of the column name), or unless you really don't care about the sort order.

- An improper comparator will sort column names inappropriately on the disk. It will be difficult (or impossible) to do range scans on column names later.

- Once defined, you can't change a comparator without rewriting all data. However, the validator can be changed later.

Design the Data Model Such that Operations are Idempotent

In an eventually consistent and fully distributed system, idempotent operations can help a lot. Idempotent operations allow partial failures in the system, as the operations can be retried safely without changing the final state of the system. In addition, idempotency can sometimes alleviate the need for strong consistency and allow you to work with eventual consistency without causing data duplication or other anomalies.

Because column family databases are fully distributed (and multi-master) in nature, write failure does not guarantee that data is not written, unlike the behavior of relational databases. In other words, even if you receive a failure for a write operation, data might be written to one of the replicas, which will eventually get propagated to all replicas. No rollback or cleanup is performed on partially written data. Thus, a perceived write failure can result in a successful write eventually. So, retries on write failure can yield unexpected results if your model isn't update idempotent.

Notes

- "Update idempotent" here means a model where operations are idempotent. An operation is called "idempotent" if it can be applied one time or multiple times with the same result.

- In most cases, idempotency won't be a concern, as writes into regular column families are always update idempotent. The exception is with the counter column family, as shown in the example below. However, sometimes your use case can model data such that write operations are not update idempotent from the use case perspective. For instance, in our earlier example, "UserbyItem" and "ItembyUser" in the final model are not update idempotent if the use case operation "user likes item" gets executed multiple times, as the timestamp might differ for each like. However, note that a specific instance of the use case operation "user likes item" is still idempotent, and so can be retried multiple times in case of failures.

Example

Suppose that we want to count the number of users who like a particular item. One way is to use the counter column family to keep count of users per item. Since the counter increment (or decrement) is not update idempotent, retry on failure could yield an over-count if the previous increment was successful on at least one node. One way to make the model update idempotent is to maintain a list of user ids instead of incrementing a count, as shown below. Whenever a user likes an item, we write that user's ID against the item; if the write fails, we can safely retry. To determine the count of all users who like an item, we read all user ids for the item and count manually.

In Figure 6-21, the update idempotent model, getting the counter value requires reading all user ids, which will not perform well (there could be millions). If reads are heavy on the counter and you can live with an approximate count, the counter column will be efficient for this use case. If needed, the counter value can be corrected periodically by counting the user IDs from the update idempotent column family.

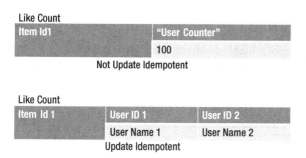

Figure 6-21. *Counter logical NoSQL data model*

Don't Use the Counter Column Family to Generate Surrogate Keys

The counter column family holds distributed counters meant for distributed counting. Don't try to use this to generate sequence numbers for surrogate keys, like Oracle sequences or auto-increment columns. You will receive duplicate sequence numbers! Most of the time you really don't need globally sequential numbers. Use "time-uuid" as surrogate keys. If you truly need a globally sequential number generator, there are a few possible mechanisms; but all will require centralized coordination and thus can impact the overall system's scalability and availability.

Favor Composite Columns over Super Columns

A super column can be used to group column keys, or to model a two-layer hierarchy. However, super columns have the following implementation issues and are therefore becoming less favorable.

Issues

- Sub-columns of a super column are not indexed. Reading one sub-column de-serializes all sub-columns.

- Built-in secondary indexing does not work with sub-columns.

- Super columns cannot encode more than two layers of hierarchy.

Similar functionality can be achieved by the use of the composite column. It's a regular column with sub-columns encoded in it. Hence, all of the benefits of regular columns, such as sorting and range scans, are available; and you can encode more than two layers of hierarchy.

For example, a composite column key like "<state|city>" will be stored ordered first by state and then by city, rather than first by city and then by state. In other words, all the cities within a state are located (grouped) on disk together.

Understanding Cassandra Data Model

The Cassandra data model is prebuilt for highly distributed and large-scale data. It trades off the traditional database guidelines (ACID compliant) to leverage operational manageability, performance, and availability.

An illustration of how a Cassandra data model would like Figure 6-22.

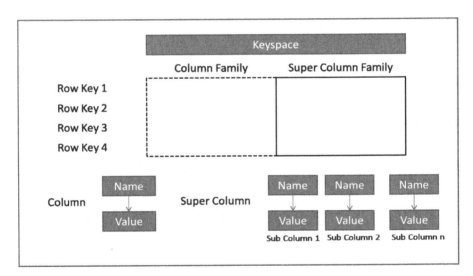

Figure 6-22. *Counter logical NoSQL data model*

The basic elements of the Cassandra data model are as follows:

- Column
- Super Column
- Column Family
- Keyspace
- Cluster

Column: A *column* is the basic unit of Cassandra data model. A column comprises name, value and a time stamp (by default). An example of column in JSON format is as follows:

```
{ // Example of Column
"name": "EmployeeID",
"value": "10029277",
"timestamp": 123456789
}
```

Super Column: A *super column* is a dictionary of boundless number of columns, identified by the column name. An example of a super column in JSON format is as follows:

```
{ // Example of Super Column
"name": "Specialization",
"value": {
"role" : {
"name": "role", "value": "Master Technology Architect", "timestamp":
123456789
},
"designation" : {
"name": "designation", "value": "Managing Director", "timestamp": 123456789
}
}
```

The major differences between a column and a super column are:

- Column's value is a string but the super column's value is a record of columns.
- A super column doesn't include any time stamp (only term's name and value).

■ **Note** Cassandra does not index sub-columns, so when a super column is loaded into memory, all of its columns are loaded as well.

Column Family (CF): A *column family* resembles an RDBMS table closely and is an ordered collection of rows, which in turn are ordered collections of columns.

A *column family* can be a "standard" or a "super" column family. A row in a *standard column family* contains collections of name/value pairs whereas the row in a super column family holds collections of super columns (group of sub-columns).

An example for a *column family* is described below (in JSON):

```
Employee = { // Employee Column Family
"10029277" : {   // Row key  for Employee ID - 10029277
// Collection of name value pairs
"EmpName" :"SM",
"mail" :"<a href="mailto:SM@xyz.com">SM@xyz.com</a>",
"phone" : "9999900000"
//There can be N number of columns
        },
"10099999" : {   // Row key  for Employee ID - 10099999
// Collection of name value pairs
"EmpName" :"MJ",
"mail" :"<a href="mailto:MJ@xyz.com">MJ@xyz.com</a>",
"phone" : "9090909090"
        },
"10199999" : {   // Row key  for Employee ID - 10199999
// Collection of name value pairs
"EmpName" :"HS",
"mail" :"<a href="mailto:HS@xyz.com">HS@xyz.com</a>",
"phone" : "9099909990"
    }
}
```

■ **Note**   Each column would contain "time stamp" by default. For easier narration, time stamp is not included here.

The address of a value in a regular column family is a row key pointing to a column name pointing to a value, while the address of a value in a column family of type "super" is a row key pointing to a column name pointing to a sub-column name pointing to a value. An example for a super column in JSON format is as follows:

```
Specialization = { // Super column family
"10029277" : {    // Row key  for Employee ID - 10029277
//Specialization skills by the employee with ID - 10029277
        "Skill1" : {"skillcode" : "skill1", "value": "BI",},
        "Skill2" : {"skillcode" : "skill2", "value": "DW",},
        "Skill3" : {"skillcode" : "skill3", "value": "CRM",},
        "Skill4" : {"skillcode" : "skill4", "value": "Analytics",},
        "Skill5" : {"skillcode" : "skill5", "value": "Big Data",}
    }
}
```

```
"10099999" : {      // Row key  for Employee ID - 10099999
//Specialization skills by the employee with ID - 10099999
    "Skill1" : {"skillcode" : "skill1", "value": "Pre-Sales",},
    "Skill2" : {"skillcode" : "skill2", "value": "BI",},
    "Skill3" : {"skillcode" : "skill3", "value": "DW",},
    "Skill4" : {"skillcode" : "skill4", "value": "Analytics",},
    "Skill5" : {"skillcode" : "skill5", "value": "Big Data",}
    }
"10199999" : {      // Row key  for Employee ID - 10199999
//Specialization skills by the employee with ID - 10199999
    "Skill1" : {"skillcode" : "skill1", "value": "Architect",},
    "Skill2" : {"skillcode" : "skill2", "value": "Data Management",},
    "Skill3" : {"skillcode" : "skill3", "value": "BI",},
    "Skill4" : {"skillcode" : "skill4", "value": "Analytics",},
    "Skill5" : {"skillcode" : "skill5", "value": "Big Data",}
    }
}
```

Columns are always organized as per the column's name within their rows. The data would be sorted as soon as it is inserted into the data model.

Keyspace: A *keyspace* is the outmost grouping for data in Cassandra, closely resembling an RDBMS database. Similar to the relational database, a keyspace has title and properties that describe the keyspace behavior. The keyspace is a container for a list of one or more column families (without any enforced association between them).

Cluster: *Cluster* is the outermost structure in Cassandra (also called as ring). The Cassandra database is specially designed to be spread across several machines functioning together that act as a single occurrence to the end user. Cassandra allocates data to nodes in the cluster by arranging them in a ring.

Table 6-2 compares the relational model with the Cassandra data model. Unlike the traditional RDBMS, Cassandra doesn't support:

- Query language like SQL (T-SQL, PL/SQL, etc.). Cassandra provides an API called "thrift" through which the data could be accessed.

- Referential Integrity (operations like cascading deletes are not available)

Table 6-2. Relational Data Model vs. Cassandra Data Model

Relational Data Model	Cassandra Data Model (Standard)	Cassandra Data Model (Super)
Server based	Cluster based	
Database	Key Space	
Table	Column Family	
Primary Key	Key	
Column Name	Column Name	Super Column Name
	Column Value	Column Value
		Column Value

Designing Cassandra Data Structures

1. **Entities and Point of Interest:** The best way to model a Cassandra data structure is to identify the entities that would be subjected to most queries and creating the entire structure around the entity. The activities performed (generally the use cases) by the user applications, how the data is retrieved and displayed would be the areas of interest for designing the Cassandra column families.

2. **De-normalization:** Normalization is the set of rules established to aid in the design of tables and their relationships in any RDBMS. The benefits of normalization would be:

   • Avoiding repetitive entries

   • Reduction of storage space

   • Prevention of schema restructuring for future needs.

   • Improved speed and flexibility of SQL queries, joins, sorts, and search results.

Achieving similar kind of performance for big data scale is a challenge in traditional relational data models. Therefore, in most of the big appl data ications de-normalization approaches are adopted to achieve performance. Cassandra does not support foreign key relationships like a relational database, and the better way is to de-normalize the data model. The important fact is that instead of modeling the data first and framing the queries, with Cassandra the queries would be modeled first and then the data be framed around them.

3. **Planning for Concurrent Writes:** In Cassandra, every row within a column family is identified by the unique row key (generally a string of unlimited length). Unlike the traditional RDBMS primary key (which enforces uniqueness), Cassandra doesn't impose uniqueness (duplicate row key insertion might disturb the existing column structure). So care must be taken to create the rows with unique row keys. Some of the ways for creating unique row keys are as follows:

- Surrogate/ UUID type of row keys

- Natural row keys

Schema Migration Approach (Using ETL)

There are various ways of migrating data from relational data structures to Cassandra structures, but if there are complex transformations and business rules involved it is always advisable to leverage a data processing layer comprising ETL utilities (Figure 6-23).

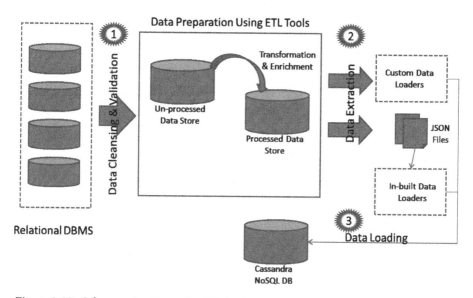

Figure 6-23. Schema migration using ETL tools

By using in-built data loaders the processed data can be extracted to flat files (in JSON format) and then uploaded to the Cassandra data structure's using these loaders. Custom loaders could be fabricated in case of additional dispensation rules, which could either deal the data from the processed store or the JSON files.

The overall migration approach would be as follows:

1. Data preparation as per the JSON file format.

2. Data extractions into flat files as per the JSON file format or extraction of data from the processed data store using custom data loaders.

3. Data loading using in-built or custom loaders into Cassandra data structure(s).

4. The various activities for all the different stages in migration are further discussed in detail in below sections.

Data Preparation and Extraction: ETL is the standard process for data extraction, transformation and loading (Figure 6-24). At the end of the ETL process, reconciliation forms an important function. This includes validation of data with the business processes. The ETL process also involves the validation and enrichment of the data before loading into staging tables.

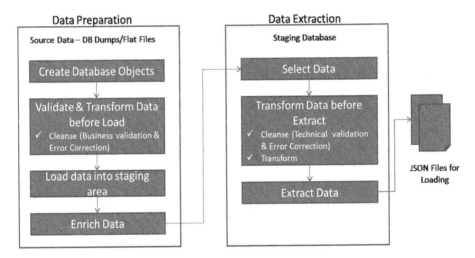

Figure 6-24. Schema migration data preparation and data extraction process

The following activities are executed during data preparation:

1. Creation of database objects: Necessary staging tables are to be created as per the requirements based on which will resemble standard open interface / base table structure.

2. Validate and transform data before load from the given source (dumps/flat files).

3. Data Cleansing:

- Filter incorrect data as per the JSON file layout specifications.

- Filter redundant data as per the JSON file layout specifications.

- Eliminate obsolete data as per the JSON file layout specifications.

4. Load data into staging area

5. Data Enrichment:

- Default incomplete data

- Derive missing data based on mapping or lookups

- Differently structured data (1 record in as-is = multiple records in to-be)

The following activities are executed during data extraction into JSON file formats:

1. Data selection as per the JSON file layout

2. Creation of SQL programs based on the JSON file layout

3. Scripts or PLSQL programs are created based on the data mapping requirements and the ETL processes. These programs serve various purposes, including the loading of data into staging tables and standard open interface tables.

4. Data transformation before extract as per the JSON files layout specification and mapping documents.

5. Flat files in form of JSON format for data loading

Data Loading: Cassandra data structures can be accessed using different programming languages like .net, Java, Python, Ruby, etc. Data can be directly loaded from the relational databases (like Access, SQL Server, Oracle, MySQL, IBM DB2, etc.) using these programing languages. Custom loaders could be used to load data into Cassandra data structure(s) based on the enactment rules, customization level, and the kind of data processing.

End Points

The traditional well-understood design approach of a data warehouse is a **central** (for Enterprise Data Warehouse) or **departmental** (for Data Marts) repository of data. ETL routines pull data from a variety of data sources, cleanse, and transform them; then they are loaded into a data warehouse. Broadly speaking, this data is organized in a dimensional data model that caters to two use-cases:

- **Canned Reports.** A set of BI queries is run with regular frequency to monitor the state of the business. Business users look at the reports to review a predefined set of business KPIs and make informed decisions.

- **Ad-hoc Queries and What-If Simulations.** Data analysts run a set of queries to analyze and find answers to specific business questions, as these questions are above and beyond the standard set of KPIs. During the course of these ad-hoc analyses the data analysts also perform several what-if scenarios while planning analysis.

The ad-hoc tasks have a defined life cycle. Once the data analysts are able to find answers to new business questions, the queries are incorporated into a report so that business users can monitor it as an ongoing practice. In a typical data warehouse, the bulk of tasks (80 percent) are from reports. The remaining 20 percent is from ad-hoc.

Since reports are frequent and generate known queries, the design of the data warehouse is done to cater to reporting. This includes data models, indexes, materialized views or derived tables (and other optimizations) to make the known reporting queries perform optimally.

Since ad-hoc tasks are infrequent and generate unknown queries, the design of the data warehouse is unable to cater to them upfront. This means that ad-hoc tasks generate queries that are harder to satisfy (since they are constrained by the data modeling decisions made for reporting) and therefore impose more load on the data warehouse.

The net result is that reports run fast while ad-hoc queries are slow. In fact, ad-hoc queries consume so much resource that reporting applications run slower, and that is not good: reports are distributed widely and reach a wide variety of business users. They are unhappy and put pressure on the data warehousing team to "get the reports in time." At the same time, the ad-hoc users are unhappy because they can't get to the data fast enough to benefit the business.

Historically, the answer to this deadlock situation was to prioritize via workload management and constrain ad-hoc usage to devote resources to reporting. If workload management didn't work, the answer was to define rules such as: *"reports will not be refreshed when data is loading"; ad-hoc queries should not be run when reports are being generated."*

Let's call this design pattern of data warehouse a **reporting data warehouse.** The priority objective of a reporting data warehouse is to ensure reports are accurate and that they perform optimally. The ad-hoc tasks are not treated with priority: they don't get dedicated data models or large chunks of resources; their tasks were heavily monitored, and often they are asked to curtail their requirements (use samples, use rolled up aggregates that were built to make reports faster, use smaller timeframes of history that were retained to just satisfy reporting requirements, phrase queries that are simpler even though they may be compromises on the pattern sought, so on and so forth).

However, with big data platforms all these constraints on ad-hoc tasks become meaningless. And with current technology advances we have the ability to address the constraints and issues discussed above.

This motivates the definition of a different design pattern of a data warehouse: a **big data and analytics data warehouse.** The priority objective of a big data analytics data warehouse is to provide capabilities for ad-hoc analytics, and the primary users of the big data and analytics data warehouse are data analysts. The platform would support schema-less data ingestion architecture that will allow the data analysts to integrate any data source into the system, the data models are built to support their ad-hoc usage: fine-granularity data is retained, rich dimension tables are frequently imported, derived views and tables are created promptly, interfaces are opened up to express their patterns in a computationally simple and natural manner, scale-out is used to create resources for the tasks to finish interactively, and enough storage is allocated for several exports to proceed simultaneously.

In summary, the design methodology of a big data and analytical data warehouse is substantially different from a reporting data warehouse. Understanding the primary customer of a data warehouse can often help simplify operations of the data warehouse and help lower the operating point costs substantially by making priorities clearer.

References

http://blogs.informatica.com/perspectives/2010/11/17/understanding-data-integration-patterns-simple-to-complex: David Linthicum, November 17, 2010

http://www.datasciencecentral.com/profiles/blogs/11-core-big-data-workload-design-patterns: Derrick Jose, August 13, 2012

http://highlyscalable.wordpress.com/2012/03/01/nosql-data-modeling-techniques: IlyaKatsov, March 1, 2012

Is Data Modeling Relevant in a NoSQL environment? Robinson Ryan

MapReduce Patterns, Algorithms, and Use Cases: Highly Scalable Blog- IlyaKatsov, February 1, 2012

http://www.codeproject.com/Articles/279947/Migration-of-Relational-Data-structure-to-Cassandra

▓ ▓ ▓

CHAPTER 7

Big Data Analytics Methodology

Big data is baffling, and analytics are complex. Together, big data analytics make a difficult and complex undertaking largely because technology architectures and methodologies are immature.

Big data analytics uncovers patterns in a wide variety of data and associates the patterns with business outcomes. Analysts use analytical techniques and tools to detect unusual, interesting, previously unknown, or new patterns in data. Big data is a result of interaction of four dimensions of scale (increasing data volumes, high velocity of data creation, increasing complexity of data types, and extreme time sensitivity of data diminishing its value if not treated at that moment) thereby posing different challenges to manage, not to mention applying analytics techniques to find new insights.

Big data does not behave the same as other data. The challenges associated with analytics on big data require a different approach from traditional data analytics processes. For example, content analysis of streaming media requires high-speed processing, storage, and fast analytics techniques. This was one of the original target applications for Google's map-reduce algorithm.

Challenges in Big Data Analysis

Heterogeneity and Incompleteness: The nuance and richness of natural language is incomprehensible. However, this is true especially in the case of big data scale where data variety is farfetched and analysis algorithms expect homogeneous data and cannot fully understand nuances. Consequently, data must be carefully structured as a first step in (or prior to) data analysis. Consider an electronic health record database design that has fields for birth date, occupation, and blood type for each patient. What do you do if a patient cannot provide one or more of these pieces of information? Obviously, the health record is still placed in the database but with the corresponding attribute values being set to "null". A data analysis that looks to classify patients by occupation, for example, must take into account patients whose occupations are not known. Worse, these patients with unknown occupations can be ignored in the analysis only if we have reason to believe that they are otherwise statistically similar to the patients with known occupation for the analysis performed.

Even after data cleaning and error correction, some incompleteness and errors in data are likely to remain. This incompleteness and these errors must be managed during data analysis. Doing this correctly is a challenge.

Scale: Managing large and rapidly increasing volumes of data has been a challenging issue for many decades. With the advent of technologies like Hadoop distributions and cloud computing, we have the ability to store massive amounts of data at relatively low cost. These innovative platforms now aggregate multiple disparate workloads with varying performance goals (e.g., interactive services demand that the data processing engine return back an answer within a fixed response time cap) into very large clusters. This level of sharing of resources on expensive and large clusters requires new ways of determining how to run and execute data processing jobs so that we can meet the goals of each workload cost-effectively, and to deal with system failures, which occur more frequently as we operate on larger and larger clusters (that are required to deal with the rapid growth in data volumes). This places a premium on declarative approaches to expressing programs: even those programs doing complex machine learning tasks, since global optimization across multiple users' programs is necessary for good overall performance. Reliance on user-driven program optimizations is likely to lead to poor cluster utilization, since users are unaware of other users' programs. System-driven holistic optimization requires programs to be sufficiently transparent, e.g., as in relational database systems, where declarative query languages are designed with this in mind.

Timeliness: The flip side of size is speed. The larger the data set to be processed, the longer it will take to analyze. The design of a system that effectively deals with size is likely also to result in a system that can process a given size of data set faster. However, it is not just this speed that is usually meant when one speaks of velocity in the context of big data.

There are many situations in which the result of the analysis is required immediately. For example, if a fraudulent credit card transaction is suspected, it should ideally be flagged before the transaction is completed, potentially preventing the transaction from taking place at all. Obviously, a full analysis of a user's purchase history is not likely to be feasible in real time. Rather, we need to develop partial results in advance so that a small amount of incremental computation with new data can be used to arrive at a quick determination.

Given a large data set, it is often necessary to find elements in it that meet a specified criterion. In the course of data analysis, this sort of search is likely to occur repeatedly. Scanning the entire data set to find suitable elements is obviously impractical. Rather, index structures are created in advance to permit finding qualifying elements quickly. The problem is that each index structure is designed to support only some classes of criteria. With new analyses desired using big data, there are new types of criteria specified, and a need to devise new index structures to support such criteria. For example, consider a traffic management system with information regarding thousands of vehicles and local hot spots on roadways. The system may need to predict potential congestion points along a route chosen by a user and then suggest alternatives. Doing so requires evaluating multiple spatial proximity queries working with the trajectories of moving objects. New index structures are required to support such queries. Designing such structures becomes particularly challenging when the data volume is growing rapidly and the queries have tight response time limits.

Privacy: The privacy of data is another huge concern, and one that increases in the context of big data. There are numerous debates regarding the inappropriate use of personal data, particularly through linking of data from multiple sources. Managing privacy is effectively both a technical and a sociological problem, which must be addressed jointly from both perspectives to realize the promise of big data.

Consider, for example, data gleaned from location-based services. These new architectures require a user to share his/her location with the service provider, resulting in obvious privacy concerns. Note that hiding the user's identity alone without hiding person's location would not properly address these privacy concerns. An attacker or a (potentially malicious) location-based server can infer the identity of the query source from its (subsequent) location information. For example, a user's location information can be tracked through several stationary connection points (e.g., cell towers). After a while, the user leaves "a trail of packet crumbs" which could be associated to a certain residence or office location and thereby used to determine the user's identity. Several other types of surprisingly private information such as health issues (e.g., presence in a cancer treatment center) or religious preferences (e.g., presence in a church) can also be revealed by just observing anonymous users' movement and usage pattern over time. In general, hiding a user location is much more challenging than hiding his/her identity. This is because with location-based services, the location of the user is needed for a successful data access or data collection, while the identity of the user is not necessary.

Human Collaboration: Despite the tremendous advances made in computational analysis, there remain many patterns that humans can easily detect but that computer algorithms have a hard time finding. Ideally, analytics for big data will not be all computational; rather these will be designed explicitly to have a human in the loop. The new sub-field of visual analytics is attempting to do this, at least with respect to the modeling and analysis phase in the pipeline. There is similar value to human input at all stages of the analysis pipeline.

In today's complex world, it often takes multiple experts from different domains to really understand what is going on. A big data analysis system must support input from multiple human experts, and shared exploration of results. These multiple experts may be separated in space and time when it is too expensive to assemble an entire team together in one room. The data system has to accept this distributed expert input and support their collaboration.

System Architecture: Business data is analyzed for many purposes: a company may perform system log analytics and social media analytics for risk assessment, customer retention, brand management, and so on. Typically, such varied tasks have been handled by separate systems, even if each system includes common steps of information extraction, data cleaning, relational-like processing (joins, group-by, aggregation), statistical and predictive modeling, and appropriate exploration and visualization tools as discussed in the methodology in this chapter.

With big data, the use of separate systems in this fashion becomes prohibitively expensive given the large size of the data sets. The expense is due not only to the cost of the systems themselves but also to the time to load the data into multiple systems. Consequently, big data has made it necessary to run heterogeneous workloads on a single infrastructure that is sufficiently flexible to handle all these workloads. The challenge here is not to build a system that is ideally suited for all processing tasks. Instead, the need is for the underlying system architecture to be flexible enough that the components built on top of it for expressing the various kinds of processing tasks can tune it to efficiently run these different workloads.

If users are to compose and build complex analytical solutions over big data, it is essential that they have appropriate high-level primitives to specify their needs in such flexible systems. The map-reduce framework has been tremendously valuable, but is only a first step. Even declarative languages that exploit it, such as Pig Latin, are at a rather low level when it comes to complex analysis tasks. At present big data analytics solutions employ a host of tools/processes to develop an end-to-end production-ready system. Each operation within the system (cleaning, extraction, modeling, etc.) potentially runs on a very large data set. Furthermore, each operation itself is sufficiently complex that there are many choices and optimizations possible in how it is implemented.

The very fact that big data analysis typically involves multiple phases highlights a challenge that arises routinely in practice: production systems must run complex analytic pipelines, or workflows, at routine intervals, e.g., hourly or daily. New data must be incrementally accounted for, taking into account the results of prior analysis and preexisting data. And of course, provenance must be preserved, and must include the phases in the analytic pipeline. Current systems offer little to no support for such big data pipelines, and this is in itself a challenging objective.

In the sections below we will discuss a methodology that outlines an approach for developing big data analytics solutions.

Big Data Analytics Methodology

The big data analytics methodology is a combination of sequential execution of tasks in certain phases and highly iterative execution steps in certain phases. Because of the scale issue associated with big data system, designers must adhere to a pragmatic approach of modifying and expanding their processes gradually across several activities as opposed to designing a system once and all keeping the end state in mind.

Figure 7-1 provides a high-level view of the big data analytics methodology, and big data analytics designers (i.e., architects, statisticians, analysts, etc.) are advised to iterate through the steps outlined in Figure 7-1. The designer should plan to complete several cycles of design and experimentation during steps 2 through 5. Each cycle should include additional and larger data samples and apply different analytics techniques as appropriate for data and relevant for solving the business problem. Designers should revisit the entire framework periodically after the system starts running in production (steps 1 through 7; see Figure 7-1).

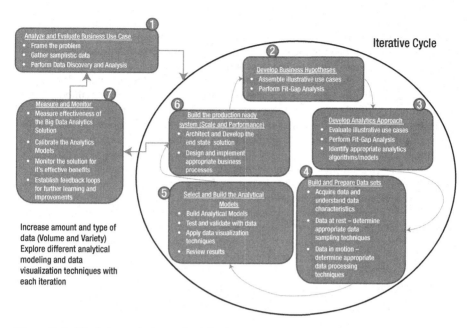

Figure 7-1. *Big data analytics methodology*

You might sense a certain amount of similarity of this methodology with other data analytics implementation and BI methodologies; however, the above methodology differs from others by the number of times the designer should execute the steps to solve design problems associated with processing at full scale. Knowledge gained during each pass through of the various steps should be reflected in the system design.

Analyze and Evaluate Business Use Case

The first step in the methodology is to analyze and evaluate the business use case. In many instances the analyze and evaluate step is also considered as a proof of concept (POC) exercise. It is not uncommon to notice the first few cycles in this step will likely produce some unexpected results, for example:

- The data samples do not include enough descriptive information to find the desired correlations.

- The data discovery and analysis activities do not scale due to the extreme number of early patterns observed in the sample data.

- The POC infrastructure is not good enough to handle the variety of data and the types of algorithms applied exhibit a good number of shortcomings.

The most important outcome of the analyze and evaluate step is the development of a detailed description of the business hypotheses inclusive of appropriate infrastructure, software and data needs. It requires a business analyst to take the business's requirements and create a comprehensive document. The document must be detailed enough for a business user to understand the business and technical criteria that the project will succeed or fail under. During this first step, you will usually have a fuzzy notion of the requirements. The architect must accept this as a limitation and devise methods to further refine the requirements till a certain acceptable level of clarity is obtained.

A properly framed requirement consists of a description of the issue the business wants to analyze, the issue's importance to the organization, collection of a set of reasons supporting the issue's creation, and analysis of any constraints such as time, place, or condition.

Consider two examples of a properly framed requirement: the loan repayment delinquency problem and the product mix optimization problem.

The Loan Repayment Delinquency Problem: The credits and loans processing department of a bank requests that an analytics team investigate a delinquency problem where increasingly a large base of their customers are turning delinquent. The business sponsor wants the analytics team to help them find the answer to the question: "Why are they seeing a disturbing trend of their customer base turning delinquent?"

After a detailed workshop with the business users, the analysts started developing the business use case:

- **Problem:** What is causing customers to turn delinquent?

- **Behavior:** Who are these customers? Are they long-time, profitable customers suddenly turning delinquent? Or are they newly acquired customers?

- **Complications:** What conditions constitute delinquency? Missing payment dates consecutively for last 3 months, missing payment dates for last 2 months but paying back the money with interest fees over the subsequent months, not paying over the previous 6 months and not reachable?

- **Impact to business:** Reducing profitability per customer, revenue loss, and increased expenses chasing the delinquent customers.

- **Background:** The bank recently launched multi-channel interactions with customers in order to acquire more customers. While the customer base increased, and a large chunk of the customers came through the online channels, there was a need to expedite the loans and credit approval processes so that the customer does not go away to other banks that had also launched similar products and interaction channels.

- **Conditions:** The first few symptoms of missing payments happen after the first 3 months of becoming a customer. There are also observations across the older customer base of older customers who are turning delinquent (although not frequently).

The delinquency problem meets the requirement for a big data problem due to volume, complexity, and variety. The bank acquires millions of customers annually. Business users believe the delinquency trend is not new and may be due to several factors: poor economy, competition luring potential customers away with the assurance that their loans from the bank will have a reduced interest fee, bank's sales agents signing up all these customers to meet their targets and earn commissions without doing a proper investigation, etc.

Product Mix Optimization Problem: The marketing department of a major consumer packaged goods company asks an analytics team to help them understand the following: Which types of product mix will yield high sales by region but also help replenish the inventory by bundling in least selling products in the mix? And how do customers feel about these product mix strategies?

- **Problem:** Which product mix bundles are most effective in each region? What are the unique constituents of these bundles (how many are big sellers and how many are marginal sellers)? What comments, if any, appear in social media, and do these comments have a positive or negative effect on sales?

- **Behavior:** Consumer goods customers buy the bundled products from outlets and write about the products in social media.

- **Complications:** The company does many promotions and advertisements through various channels; these promotional activities are not tracked to purchases. What behaviors constitute an effective product mix strategy? Does the product mix bundle lead to repeat sales? How do consumers feel about the product mix and the value?

- **Impact:** Increased revenue from sales of product mix; increased sell of previously least selling products.

- **Background:** The marketing department started advertising through different media channels and different regions. The company used weekend newspaper advertisements in their eastern and southern regions, a push mail mechanism was used with a select set of customers in the western region, web ads in all regions, outbound call center conversations targeting northern regions. The vice president of sales noted receipt of "dozens of comments" on his Facebook page.

- **Conditions:** Product sales seem to increase following the various product mix bundles that the company used to do, but the conditions to support such a hypothesis are unknown. Marketing does not know if the social media information has any effect on sales or not.

The product mix optimization problem qualifies as a big data problem across the four dimensions of big data. The consumer products company pays heavy fees to subscribe to point of sale (POS) data from its outlets. POS data is highly complicated and is ridden with a multiplicity of codes for each of the company's numerous products.

The outlets do not have a standardized data structure and often use a large number of rules; plus, much of it is unstructured and not documented properly. Day of the week and seasonality may also be among the conditions precedent to successful and unsuccessful sales. Finally, social media data varies greatly by source, interaction among groups, and the semantics that express sentiments concerning the promotion and the product mix.

Develop Business Hypotheses

Should it be the availability and richness of data that decide the scope of business requirements and expectations, or should it be the other way around? How to proceed then?

As we discussed above, the two examples require the analyst to not only understand the business problem but also several factors influencing the behaviors (e.g., either customers turning delinquent or product mix bundles sales); but the outcome of the analysis cannot be known until the system is built. Often, the designer must make a best guess as to the outcome of an analysis, perform data exploratory analysis by looking into all possible data sources as relevant for solving the business problem, and then perform analytics. With preliminary results in hand, business users then can respond with more certainty as to the sufficiency and usefulness of the results. It is recommended to continue the iterations until business users and designers agree upon the business and technical requirements.

Examples of Business Hypotheses

We will use the two analytics problems cited above to illustrate the second step of the methodology—develop the business hypotheses.

Loan Repayment Delinquency Problem: A customer profile inclusive of transactions can run into thousands of attributes. To properly understand which attributes are key to solving the problem, the analyst will have to define the limits. The analyst further explores and discovers that fewer than a hundred delinquencies per type of customer account carry insufficient value to be of interest to the company. Therefore, the analyst sets a rule equal to or greater than 100 incidents against those types of accounts, thereby excluding inconsequential results. The business users then explain to the analysts that they expect the delinquency behaviors are observed after a sequence of transactions in certain age groups that were not observed in earlier delinquent behaviors, and these behaviors are recent. Based on these additional inputs from business users, the analysts now can start creating different sample sizes—a random selection of records from the customer base.

Product Mix Optimization Problem: The marketing department staff provides the analysts with its best guess for the product mix bundling. They form product mix bundles by understanding customer buying behavior patterns, seasonality, and special occasions such as outlet's anniversary, etc. Moreover, these product mix bundles are independent of the channel through which the customers receive the promotion. The analyst prepares a data set that shows the product mix bundle across several categories of products. Against each of these product mix bundles, there is a sell value, highest applicable discount percentage, the effective dates, listing of best-selling products in the bundle, and a listing of the lowest-selling products in the bundle, and a brief description of the product mix bundle.

Develop the Analytics Approach

The analytics approach defines the analytical techniques the analyst will use to solve business problem. The solution to a particular problem requires the application of the correct analytical method and data.

The loan delinquency problem requires the analysts to explore several data points to determine if a particular pattern or sequence of events is the root cause of the observed pattern. The analyst considers the data collected across different segments of the customer base. For instance, it was observed that, in general, customers classified as high net individuals do not exhibit delinquency, but recently classified customers as high net individuals who also opened demat (internet-based) accounts less than 6 months ago are exhibiting delinquency in their credit card payments schedules. In this manner, the analyst can keep on exploring the relationship that exists between several components; however, this would become cumbersome because of the combinatorial complexity of the analysis.

The product mix optimization appeal problem requires marketing to track sales by region, product mix type, and influence by social media comments, if any. The analyst decides to explore through several data points to arrive at increased sales patterns.

Note However, there is a broader question that also needs to be addressed. Is the problem a simulation or a dynamic modeling issue?

For example:

- Does the organization require an econometric model to determine whether a certain combination of products in the product mix bundle and the resulting price appeals to a specific type of customer demographics? These and similar problems require knowledge of statistical analysis and data architectures to design the information models that the solution will require.

- Does the problem require an optimization program such as finding the most appealing price to put on a product mix bundle that will ensure the profit margin per product mix bundle does not drop below 10 percent? Such problems require linear-programming models that employ optimization methods such as the Simplex method.

- Does the problem require the simulation of a customer behavior right from the day they are acquired, how they were acquired, what transactions they are performing and when, across several dimensions of time and business events? Such problems require knowledge of business processes and graph theory.

- Finally, does the problem relate to a random process? For example, business users might want to determine which channels are bringing in profitable customers. The business users might also want to put in alert based controls in the system to detect delinquency behaviors as early as possible. These situations require the application of statistical analysis and statistical process modeling.

Choose the Correct Analytical Method

The analyst then decides on the types of analytics techniques to solve the business problem:

- **Validation and comparison:** In many cases just by validating and comparing behaviors across similar entities (similar type of product mix bundles, similar type of customer accounts) over a large set of records throws up interesting patterns.

- **Aggregation and summarization:** Often times the data sets may contain extreme values, null values, etc. Applying summary statistics and aggregations such as averages, standard deviations, ranges, maximums, and minimums helps in discarding data that is not appropriate for analytics and may distort the final outcomes.

- **Maximization and minimization:** The expected outcome of the business problem often falls under these two categories, either to minimize loss (loan delinquency problem) or maximize profit (product mix optimization problem).

- **Rare-event detection and unusual patterns detection:** If you are trying to find the needle in the haystack, looking for rare events that are a small fraction of the overall larger data or looking for unusual patterns across a series of seemingly valid events, then the work becomes increasingly tough. No single analytical technique or sophisticated algorithm will provide a silver bullet answer. In such cases, it is recommend applying analytics-pipeline approach and running through random data samples till you observe any meaningful patterns.

Analysis Outcomes

The real users of the analytics outcomes are business users but these users often do not understand the complex mathematical formulae, statistical analysis models, etc. Hence it is extremely important to equip the business users with easy-to-understand and highly intuitive tools through which they will understand what actions are to be performed.

- **Reports:** The business may want the analytics results to be delivered through a set of reports and dashboards highlighting the root causes and recommendations.

- **Alerts:** Oftentimes the business ask is to become predictive and apply alerts to events that exhibit fraudulent behavior, early warning for a negatively impacting outcome, highlighting a profitable business opportunity during early stages, etc.

- **Process optimization:** If you are analyzing business processes to improvise that can produce increasing business opportunities or minimize losses, the analytics outcome needs to be embedded into those process chains appropriately.

- **Model:** In cases where you are extensively doing simulations and what-if scenarios, the analytics outcome is actually a model or an application that the business users will use regularly.

Build and Prepare Data Sets

Analytics is all about developing data sets that capture not only transactional data but also interaction data depicting the inter-relationships between data due to the business events and associated context. Heterogeneity, scale, timeliness, complexity, and privacy problems with big data impede progress at all phases that can create value from data.

The problems start right away during data acquisition: you have to decide what data to keep and what to discard and how to store what you keep reliably with the right metadata. Much data today is not originally in structured format; for example, tweets and blogs are weakly structured pieces of text, while images and videos are structured for storage and display (but not for semantic content and search). Transforming such content into a structured format for later analysis is a major challenge. The value of data increases exponentially when it can be linked with other data, thus data integration is a major creator of value. Since most data is directly generated in digital format today, we have the opportunity and the challenge both to influence the creation to facilitate later linkage and to automatically link previously created data.

The most critical activity during this phase is to ascertain the completeness and richness of data. The patterns represented by the data must be random, reliable, and consistent. Randomness assures that data samples represent the statistical characteristics of the complete data set. Reliable means that analyses are sufficiently accurate for their intended business purposes. Consistent means that the same analyses of a different random sample of data from the same data source will yield the same analytical results within an acceptable error margin.

Big data does not arise out of a vacuum; it is recorded from some data-generating source. For example, consider our ability to sense and observe the world around us: from the heart rate of an elderly citizen and the presence of toxins in the air, to the planned square kilometer array telescope, which will produce up to 1 million terabytes of raw data per day. Similarly, scientific experiments and simulations can easily produce petabytes of data today.

Much of this data is of no interest, and it can be filtered and compressed by orders of magnitude. One challenge is to define these filters in such a way that they do not discard useful information. For example, suppose one sensor reading differs substantially from the rest: it is likely to be due to the sensor being faulty, but how can we be sure that it is not an artifact that deserves attention? In addition, the data collected by these sensors most often are spatially and temporally correlated (e.g., traffic sensors on the same road segment). We need research in the science of data reduction that can intelligently process this raw data to a size that its users can handle while not missing the needle in the haystack. Furthermore, we require "on-line" analysis techniques that can process such streaming data on the fly, since we cannot afford to store first and reduce afterward.

The second big challenge is to automatically generate the right metadata to describe what data is recorded and how it is recorded and measured. For example, in scientific experiments, considerable detail regarding specific experimental conditions and procedures may be required to be able to interpret the results correctly, and it is important that such metadata be recorded with observational data. Metadata acquisition systems can minimize the human burden in recording metadata. Another important issue here is data provenance. Recording information about the data at its birth is not useful unless this information can be interpreted and carried along through the data analysis pipeline. For example, a processing error at one step can render subsequent analysis useless; with suitable provenance, we can easily identify all subsequent processing that dependent on this step. Thus we need research both into generating suitable metadata and into data systems that carry the provenance of data and its metadata through data analysis pipelines.

Frequently, the information collected will not be in a format ready for analysis. For example, consider the collection of electronic health records in a hospital, comprising transcribed dictations from several physicians, structured data from sensors and measurements (possibly with some associated uncertainty), and image data such as X-rays. We cannot leave the data in this form and still effectively analyze it. Rather, we require an information extraction process that pulls out the required information from the underlying sources and expresses it in a structured form suitable for analysis. Doing this correctly and completely is a continuing technical challenge. Note that this data also includes images and will in the future include video; such extraction is often highly application dependent (e.g., what you want to pull out of an MRI is very different from what you would pull out of a picture of the stars or a surveillance photo). In addition, due to the ubiquity of surveillance cameras and popularity of GPS-enabled mobile phones, cameras, and other portable devices, rich and high-fidelity location and trajectory (i.e., movement in space) data can also be extracted.

We are used to thinking of big data as always telling us the truth, but this is actually far from reality. For example, patients may choose to hide risky behavior, and caregivers may sometimes misdiagnose a condition; patients may also inaccurately recall the name of a drug or even forget that they ever took it, leading to missing information in (the history portion of) their medical record. Existing work on data cleaning assumes well-recognized constraints on valid data or well-understood error models. For many emerging big data domains these do not exist.

Given the heterogeneity of the data, it is not enough merely to record it and throw it into a repository. Consider, for example, data from a range of scientific experiments. If we just have a bunch of data sets in a repository, it is unlikely anyone will ever be able to find, let alone reuse, any of this data. With adequate metadata, there is some hope; but even so, challenges will remain due to differences in experimental details and in data record structure.

Data for the Loan Repayment Delinquency Problem: The analyst scouts the various data systems in the bank for correct and relevant data:

- Customer accounts data that includes when it was created, how it was created, and who created it

- Customer transactions data to understand the transaction behaviors of entire population

- Customer interaction channels data to understand what is influencing inflow of customers and how reliable/profitable those channels are

- Bank customer agents' performance data to detect any correlation between customer delinquency and the agents involved in creating the accounts

- External data such as credit bureau data for credit scores of customers

- External data such as customer life style data and what influence it has on delinquency

Data for the Production Mix Optimization Problem: The analyst provisions data as discussed below:

- Sales data that includes product bundles, product bundle constituents, store, location, and seasonality information

- Customer information from the sales system enhanced with demographic data to better understand those customers

- Time and date information with definitions of seasons

- Product cost information to calculate profitability

After designing the data architecture and developing the data sets, the analyst is now ready to run appropriate analytics models on these data sets. The goal of this step in the methodology is to arrive at root causes or detect patterns.

Select and Build the Analytical Models

Analytics models can become quite complex; hence, considerations should be made to infuse flexibility, as determining appropriate data samples and selecting appropriate algorithms are key to moving ahead. Especially in the case of big data scenarios, it is always seen a prototype developed with a selective set of analytics algorithms that uses small but meaningful data samples. The results of experiments using a prototype helps the analyst to further refine the algorithms or choose a different algorithm as well as enrich the current data sets or add new data sets till the desired outcome is achieved. All of the above steps are performed in a heavily iterative fashion, with every iteration constantly improving on analysis results and often seeking additional data to enrich the analysis process (see Figure 7-2).

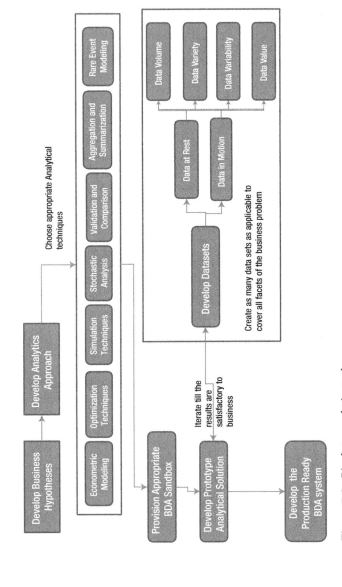

Figure 7-2. *Big data analytics tasks*

Often, a data analysts use industry accepted analytics products like SAS, R, SPSS, SQL Server Analytics Services, Oracle Data Miner, Microsoft Excel, or other analytics tools.

Design for Big Data Scale

Once the prototype results are evaluated, found to be of high quality, and meet the business expectations, the next step is to build an analytics system that will handle the scale of big data challenges with desired performance levels. Not every analytical method that works with small amounts of data will scale to big data problems. Therefore, the analyst must choose a method that will scale to big data production size. There are several considerations in designing an analytics system for production readiness:

- **Complexity:** How complex are the data sets, and how complex are the analytics algorithms? Do they require special provisioning and skills to manage?

- **Efficiency:** How efficient is the analytical model? Does it require specialized s/w and h/w configurations?

- **Performance:** How capable is the analytics algorithm of running across big data dimensions? Does it require in-database processing? Does it require in-memory processing? Does it have the ability to run massively parallel processing across huge data sets and grid architectures?

- **Reliability and Accuracy:** How calibrated is the analytics output? What are the confidence intervals and error measures? Does it throw false positives and false negatives within an acceptable range?

- **Coverage and Reach:** Does the system have the ability to cover all data types and to do depth-search as well as breadth-search across several data dimensions?

- **Flexibility:** How flexible is the analytics system in adopting new algorithms?

Build the Production Ready System

Usually there is a combination of two approaches, data exploratory or analytical-pipeline processing, that are applied when building production-ready analytics systems.

Analysts are often not sure what to look for. In essence it is a highly exploratory approach they take to look at all possible data sources and relate these to a particular business problem. In such scenarios, instead of taking the entire dump of those data sources, they revert to data-sampling techniques; this approach reduces the scale of the big data along one or more of its dimensions while still faithfully representing the characteristics of the original data: i.e., the data itself, the information content represented by the data, or the information content and the data taken together. However, data samples are valid for analysis if and only if patterns in the data remain stable during the entire phase of analysis.

Analytical-pipeline is a highly automated processing architecture pattern in which sets of homogenous data are exposed to one or more analytical algorithms and techniques (refer to analytical techniques in Figure 7-2). The main objective of the analytical pipeline processing approach is to process each data set through a series of steps, preferably only once: once the data set is processed the results are then analyzed for their possible relevance to the business problem (see Figure 7-3).

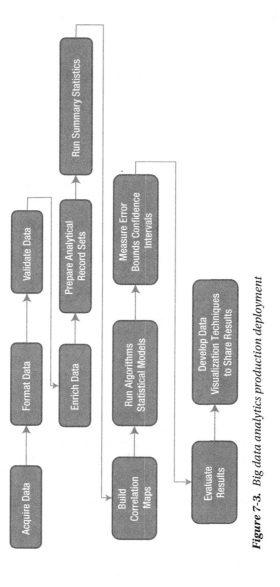

Figure 7-3. *Big data analytics production deployment*

Analytical-pipeline processing architecture is essential for big data scenarios when we're dealing with large-scale volume of data or velocity dimensions of data. Analytical techniques that explore correlations among variety of data also benefit from this approach.

Certain big data analytics use cases call for a combination of data exploration and analytical pipeline processing. For such cases, the data exploration process must be done first, and then the finalized data sets are fed into the analytical pipeline processing steps.

■ **Note** The data exploration approach is recommended for analyzing big data when the business problem is exploratory in nature, and patterns in the big data are not high- velocity driven:

- Understand several data sources before it is finalized

- The business problem falls under rare-event analysis or anomaly detection

Follow data patterns and behaviors exhibited to establish audit trails for regulatory compliance

Analytical pipeline processing is recommended for analyzing big data when the business problem is to look for definitive outcome, and patterns in the big data are at scale covering both data at rest and data in motion:

- Large-scale dimensions of big data are involved such as volume, variety, and velocity.

- The business problem falls under maximization or minimization scenarios.

Setting up the Big Data Analytics System

One of the most challenging tasks when getting started with a big data analytics system (Hybrid architectures consisting of Hadoop ecosystem, RDBMS, data visualization tools, analytics tools, and data management processes) is figuring out how to take the tools you have and put them together. The Hadoop ecosystem encompasses about a dozen different open-source projects. How do we pick the right tools for the job?

In most of the data-processing and analysis types of projects you will find three components that help in establishing an end-to-end data pipeline (from raw data to insight generation). These components are data ingestion, data store, and data analysis. A data ingestion system is the connection between the source systems and the data store where the acquired data will reside. A data analysis system is used to process the data and produce actionable insights.

Let's work through an example application and use Flume, HDFS, Oozie, and Hive to design an end-to-end data pipeline that will enable us to analyze Twitter data.

Example Application: Measuring Influence

Social media (especially Twitter) has truly become a rich data source for marketing teams. Twitter provides a powerful interaction platform to engage users. Among other things they can share with each other, users also provide word-of-mouth views on various topics. Marketing teams have always struggled to find the key influencers in a group that if targeted, can actually lead to a potentially larger audience, and the twitter platform provides enough clues to solve this problem.

Let's first understand the mechanics of Twitter. In Twitter parlance, a user (let's call him "HS") follows a set of people and has a set of followers. When HS sends an update, that update is seen by all of his followers. HS can also re-tweet other users' updates. A re-tweet is a repost of an update, much like you might forward a message or an e-mail. If MJ sees a tweet from HS, and re-tweets it, all of MJ's followers see HS's tweet, even if they don't follow HS. Through re-tweets, messages get passed much farther than just the followers of the person who sent the original tweet.

Thus, from a target base perspective, it is critical to know and engage with users whose updates tend to generate lots of re-tweets. Since Twitter tracks re-tweet counts for all tweets, we can find the users who are the leaders and who are the followers.

Now we know the question we want to ask: Which Twitter users get the most re-tweets? Who is influential within the network?

How do you answer these questions?

Your task is to find out which users are responsible for the most re-tweets. Twitter streaming API outputs tweets in a JSON format, which can be very complex. Storing this data in a traditional RDBMS will be difficult, and certainly querying this data from RDBMS system will be highly cumbersome. In the Hadoop ecosystem, the Hive component acts as a data warehouse environment for HDFS, it also provides a query interface that can be used to query data that resides in HDFS. The query language looks very similar to SQL.

So, how do you get twitter data into Hive?

Figure 7-4 shows a data flow view of how we can get data from twitter into the Hadoop ecosystem.

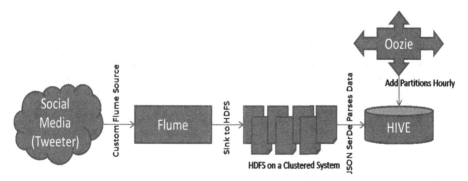

Figure 7-4. *Big data analytics system for processing tweeter data*

215

Gathering Data with Apache Flume

The twitter streaming API provides a constant stream of tweets. To gather the feeds, one option would be to use a simple utility like *curl* to access the API and then periodically load the files. However, this would require us to write code to control where the data goes in HDFS. The second option will be to use specialized components like Flume within the Hadoop ecosystem to automatically move the files from the API to HDFS, without manual intervention.

Flume is a data ingestion utility that is configured by defining endpoints in a data flow called "sources and sinks". In Flume, each individual piece of data (tweets, in our case) is called an event; sources produce events and send the events through a channel, which connects the source to the sink. The sink then writes the events out to a predefined location. For our use case, we'll need to design a custom source that accesses the twitter-streaming API and sends the tweets through a channel to a sink that writes to HDFS files. Additionally, we can use the custom source to filter the tweets on a set of keywords to help identify relevant tweets.

Partition Management with Apache Oozie

Once we have the twitter data loaded into HDFS, we can stage it for querying by creating an external table in Hive. Using an external table will allow us to query the table without moving the data from the location where it ends up in HDFS. To ensure scalability, as we add more and more data, we'll need to also partition the table. A partitioned table allows us to prune the files that we read when querying, which results in better performance when dealing with large data sets. However, the Twitter API will continue to stream tweets, and Flume will perpetually create new files. We can automate the periodic process of adding partitions to our table as the new data comes in.

Apache Oozie is a workflow coordination system that can be used to solve this problem. Oozie is an extremely flexible system for designing job workflows and can be scheduled to run based on a set of criteria. We can configure the workflow to run an ALTER TABLE command that adds a partition containing the last hour's worth of data into Hive, and we can instruct the workflow to occur every hour. This will ensure that we're always looking at up-to-date data.

Querying Complex Data with Hive

Before we can query the data, we need to ensure that the Hive table can properly interpret the JSON data. By default, Hive expects that input files use a delimited row format, but our Twitter data is in a JSON format, which will not work with the default settings. This is actually one of Hive's biggest strengths. Hive allows us to flexibly define and redefine how the data is represented on disk. The schema is only really enforced when we read the data, and we can use the Hive SerDe interface to specify how to interpret what we've loaded.

SerDe stands for "serializer" and "deserializer", which are interfaces that tell Hive how it should translate the data into something that Hive can process. In particular, the deserializer interface is used when we read data from disk, and converts the data into objects that Hive knows how to manipulate. We can write a custom SerDe that reads the JSON data in and translates the objects for Hive.

The SerDe will take a tweet in JSON form, and translate the JSON entities into columns:

```
SELECT created_at, entities, text, user
FROM tweets
WHERE user.screen_name='HS'
  AND re-tweeted_status.user.screen_name='MJ';
```

See results in Table 7-1.

Table 7-1. Results of the SerDe Function on the JSON Entity

Created At	Entities	Text	User
Mon Apr 29 13:58:23 +0000 2013	{ "urls": [], "user_mentions": [{"screen name": "HS", "name": "Harsha Srivatsa"}], "hashtags": [{"text": "BigDataAnalytics"}]}	RT@HS: #BigDataAnalytics – It is not bigness of big data that is interesting, it is the value that you can derive from all these data that can make huge business impacts	{"screen name": "MJ", "name": "Madhu Jagadeesh", "friends_count": 176, "followers_count": 231, "statuses_count": 2458, "verified": "false", "utc_offset": null, "time_zone": null}

We've now managed to put together an end-to-end system, which gathers data from the twitter-streaming API, sends the tweets to files on HDFS through Flume, and uses Oozie to periodically load the files into Hive, where we can query the raw JSON data, through the use of a Hive SerDe.

The tweeter data has some structure, but certain fields may or may not exist. The re-tweeted_status field, for example, will only be present if the tweet was a re-tweet. Additionally, some of the fields may be arbitrarily complex. The hashtags field is an array of all the hashtags present in the tweets, but most RDBMSs do not support arrays as a column type. This semi-structured quality of the data makes the data very difficult to query in a traditional RDBMS. Hive can handle this data much more gracefully.

The query below will find usernames, and the number of re-tweets they have generated across all the tweets that we have data for:

```
SELECT
t.re-tweeted_screen_name,
sum(re-tweets) AS total_re-tweets,
count(*) AS tweet_count
```

```
FROM (SELECT
re-tweeted_status.user.screen_name as re-tweeted_screen_name,
re-tweeted_status.text,
max(re-tweet_count) as re-tweets
    FROM tweets
    GROUP BY re-tweeted_status.user.screen_name,
re-tweeted_status.text) t
GROUP BY t.re-tweeted_screen_name
ORDER BY total_re-tweets DESC
LIMIT 10;
```

The result of this query could be similar to Table 7-2.

Table 7-2. *Results of tweets count*

Re-tweeted_screen_name	Total_re-tweets	Tweet_count
MJ	421	5
SS	324	7
SK	213	12
SM	199	23
JM	287	21
AB	263	15
KA	195	18
DW	86	4
AR	67	6
SP	372	29

From these results, we can see whose tweets are getting seen by the widest audience and also determine whether these people are communicating on a regular basis or not. We can use this information to carefully target our messaging.

Measure and Monitor

Having the ability to analyze big data is of limited value if users cannot understand the analysis. Ultimately, a decision maker, provided with the result of analysis, has to interpret these results. This interpretation cannot happen in a vacuum. Usually, it involves examining all the assumptions made and retracing the analysis. Furthermore, there are many possible sources of error: computer systems can have bugs, models almost always have assumptions, and results can be based on erroneous data. For all of these reasons, no responsible user will cede authority to the computer system. Rather the analyst will try to understand, and verify, the results produced by the computer.

This is particularly a challenge with big data due to its complexity. There are often crucial assumptions behind the data recorded. Analytical pipelines can often involve multiple steps, again with assumptions built in.

In short, it is rarely enough to provide just the results. Rather, one must provide supplementary information that explains how each result was derived and based upon precisely what inputs. Such supplementary information is called the "provenance" of the (result) data. By studying how best to capture, store, and query provenance, in conjunction with techniques to capture adequate metadata, the analytics system should have tools to provide users with the ability both to interpret analytical results obtained and to repeat the analysis with different assumptions, parameters, or data sets.

Tools with a rich palette of visualizations become important in conveying to the users the results of the queries in a way that is best understood in the particular domain. Whereas early business intelligence systems' users were content with tabular presentations, today's analysts need to pack and present results in powerful visualizations that assist interpretation and support user collaboration.

Furthermore, with a few clicks the user should be able to drill down into each piece of data that they see and understand its provenance, which is a key feature to understanding the data. That is, users need to be able to see not just the results but to also understand why they are seeing those results. However, raw provenance, particularly regarding the phases in the analytics pipeline, is likely to be too technical for many users to grasp completely. One alternative is to enable the users to "play" with the steps in the analysis and make small changes to the pipeline, for example, or modify values for some parameters. The users can then view the results of these incremental changes. By these means, users can develop an intuitive feeling for the analysis and also verify that it performs as expected in borderline cases. Accomplishing this requires the system to provide convenient facilities for the user to not only measure but also monitor and review results.

Establish a Support Team

Before putting the analytics models in production systems, the analytics project lead should form a specialized task force to support the business users. The team should consist of the following roles:

- A combination of business users and technical personnel who can manage and monitor the analytics models performance.

- Develop a communication plan and escalation path to resolve conflicts as they arise, as issues can happen due to quality of data, relevance of data sources, performance of the algorithms, calibration of results as more and new data are exposed to the analytics models, etc.

It is not uncommon to see analytics modules producing many more patterns than a business can use. Therefore, the analysts and the business users should evaluate the results, pursue the outcomes that show significant benefits, and discard the ones that do not.

During the measure and monitor phase of the methodology, analysts should document the process outcomes and analytical results toimprove the current system and to assist with future endeavors.

End Points

We have entered an era of big data. Through better analysis of the large volumes of data that are becoming available, there is the potential for making faster advances in many scientific disciplines and improving the profitability and success of many enterprises. However, many technical challenges described earlier in this chapter must be addressed before this potential can be fully realized. The challenges include not just the obvious issues of scale but also heterogeneity, lack of structure, error handling, privacy, timeliness, provenance, and visualization—at all stages of the methodology from data acquisition to result interpretation. These technical challenges are common across a large variety of application domains, and therefore they are not cost effective to address in the context of one domain alone.

Big data has to be managed in context, which may be noisy, heterogeneous, and might not include an upfront model. Doing so raises the need to track provenance and to handle uncertainty and error: topics that are crucial to success, and yet rarely mentioned in the same breath as big data. Similarly, as far as considering the questions related to the appropriate business problem definition, choosing the right analytics approach will typically not be laid out in advance. You may need to figure out good questions based on the data.

Fortunately, existing computational techniques can be applied, either as is or with some extensions, to at least some aspects of the big data problem. For example, relational databases rely on the notion of logical data independence: users can think about what they want to compute, while the system (with skilled engineers designing those systems) determines how to compute it efficiently. Similarly, the SQL standard and the relational data model provide a powerful uniform language to express many query needs and, in principle, allow customers to choose between vendors, thereby increasing competition. The challenge ahead is to combine these healthy features of prior systems and devise novel solutions to the many new challenges of big data.

References

BDBA: A Framework for Big Data Behavioral Analytics: Gartner
Big Data: Hadoop, Business Analytics and Beyond: A Big Data Manifesto from the Wikibon
    Community – Jeff Kelly
http://www.infoq.com/articles/BigDataBlueprint

CHAPTER 8

■ ■ ■

Extracting Value From Big Data: In-Memory Solutions, Real Time Analytics, And Recommendation Systems

Data is everywhere, but few organizations are deriving the full value from their data. How do you keep up with the velocity and variety of data streaming in and get actionable insights from it, all in real time?

The main driver for many of the innovations around big data analytics is "time to action." Today, every business is transforming itself into a *digital business.* The resulting effect is proliferation of online applications, social-mobile applications, and SaaS applications. This changing nature of business has brought the demand for real-time analytics to mainstream business approaches. There are several examples of innovative applications and usage of real-time analytics:

- A financial lending application reviewing incoming credit applications and apportioning funds across these requests to continuously minimize overall credit risk.

- An e-commerce application scanning shopping carts to detect popular product categories and optimize offers on the website in real time.

- A fraud detection system, for credit card fraud transactions, analyzing a flow of transactions to detect potential fraud and quickly allocating resources to the highest threats.

- A logistics system or real-time control system (such as a smart grid) watching changes to assets within the system and alerting when potentially dangerous conditions are detected.

- An asset management system, continuously polling machine generated data emitted through sensors, analyzing them in real-time and raising alerts if pre-defined levels of controls are violated.

221

In the earlier chapters we discussed how newer infrastructures and technologies like Hadoop, NoSQL, and parallel processing platforms are solving the challenges of processing massive amounts of data in a shorter time and with lower cost. Now, Hadoop has almost become the de facto standard for many of the batch processing analytics applications. While Hadoop and map-reduce in general do a pretty good job in processing massive amounts of data through parallel batch processing, they weren't designed to serve the real-time part of the business.

Before we deep dive into architectural constructs and discuss solutions, let's understand few key concepts.

In-Memory Database Grids. In-memory data grids were originally designed to complement traditional databases by allowing critical pieces of fast-changing data and application logic to operate at the memory layer with much higher throughput and lower latency. An in-memory database grid stores data as objects in memory, avoiding expensive disk round trips. The data model is usually object-oriented (serialized) and non-relational, organized as collections of logically related objects that can be rapidly created, updated, read, and removed. A common implementation scenario of in-memory data grid is as a "distributed cache" for one or more databases. The in-memory data grids are built on Java, allowing the grid to run embedded inside the application server cluster eliminating the traffic to the database servers.

It is not a new attempt to use main memory as a storage area instead of a disk. There are numerous examples of effectively using main memory databases, as they perform much faster than disk-based databases. When you SMS or call someone, most mobile service providers use main memory database to get the information about your contact as soon as possible. The software on your cell phone also uses main memory database effectively to show the caller details including the picture.

There are many in-memory data grid products, both commercial and open source. Some of the most commonly used products are Oracle Coherence, IBM Websphere eXtreme Scale, Hazelcast, JBoss Infinispan, GridGain, DataGrid, VMware Gemfire, Oracle Coherence, Gigaspaces XAP, Terracotta Ehcache and BigMemory.

Distributed caching products like Memcached provide a simple, high performance, in-memory key-value store. Its "scalability" is addressed through making servers completely independent of each other. The client (configured with a list of all servers) ties all the data in the servers together. A hash function maps the keys to servers on each client, thus ensuring consistency of data even in the case where all clients to have identical server lists. Data consistency becomes a concern when different clients have different server lists or different hash functions. Distributed caching do not have built-in support for replication and no native support for high availability, so any network partition or server crash leads to a loss of availability.

In contrast, in-memory data grids are fully clustered and are always aware of each other. They use a variety of algorithms to establish distributed consensus and ensure higher levels of consistency guarantees. In addition in-memory data grids provide support for distributed transactions, scatter-gather parallel query processing, tiered caching, publish-subscribe event processing, a framework to integrate data with existing databases, replication over wide area networks, etc.

In-memory data grids also enable new computing paradigms for cloud, complex event processing and data analysis. Cloud deployments promise dynamic scalability irrespective of the spikes in capacity. When spikes occur, the automatic detection and provisioning of resources (h/w capacity) is handled using virtualization. In-memory data grid, in such cases can elastically expand or contracts without any operator intervention. In-memory data grids offer complex event processing through a feature called "continuous querying." A common scenario would be, for frequently accessed data sets or event status in operational systems, the in-memory data grid can schedule the queries to be running continuously, as and when the query result set is impacted due to updates the in-memory data grid can asynchronously push "change events" to receiving applications. This feature of in-memory data grid is influencing creation of a new breed of real-time, push-oriented applications where events can be pushed all the way to thousands of devices running applications.

In-Memory Analytics. In-memory analytics is an approach to querying data when it resides in a computer's random access memory (RAM), as opposed to querying data stored on physical disks. This results in vastly shortened query response times, allowing BI and analytic applications to support faster business decisions.

As the cost of RAM declines, in-memory analytics is becoming feasible for many businesses. BI and analytic applications have long supported caching data in RAM, but older 32-bit operating systems provided only 4 GB of addressable memory. Newer 64-bit operating systems, with up to one terabyte (TB) addressable memory (and perhaps more in the future), have made it possible to cache large volumes of data potentially an entire data warehouse or data mart in a computer's RAM.

In-Memory Computing Technology: Guidelines

Are there any rules, best practices, or guidelines regarding "analytical problems, specialized use cases, or architecture related scenarios better suited for in-memory technologies"? There are a few factors, which can let you know when to go for in-memory technologies:

- There are types of workloads that need repetitive access to the entire data set and subsequent processing of the data set. You may want to keep drilling down on the result set to your original query. In addition, you may want to keep trying new visualizations of substantially the same data set. In such scenarios you shouldn't be going to the disk each time.

- Some algorithms rely on fairly random access to the data. Tin this case, it's best to keep the whole data set in memory. In particular, approaches to relationship analysis, graph processing, depth and width based searches, and experimentation or discovery type tasks tend to be fundamentally in-memory. In addition, some workloads like complex event processing need such low latency that there's no time to write the data to disk before analyzing it.

- There are some products (especially ERP, SCM, and CRM packaged applications), where the in-memory data stores and processing technologies are packaged together providing better options to store, manage, and process data faster.

Key question is: what kind of workload, in principle, should be done in memory?

Let's start by looking at some scenarios where in-memory is not only preferred but also necessary:

- **Your database is too slow for interactive analytics.** *Not all databases are as fast as we would like them to be.* This is especially true for online transaction processing (OLTP) databases that are meant to store transactional data. If you are working with a slow database, then you may want to move your data in-memory, so you can perform interactive, speed-of-thought analysis without being constantly slowed down waiting for queries to return result sets from disks.

- **You need to take load off a transactional database.** Regardless of the speed of your database, when its primary purpose is storing and processing transactional data, you don't want to put additional load on it. Analytical queries can put tremendous pressure on transactional database and slow it down, negatively impacting mission critical business operations. Bringing in a set of data to an in-memory space increases the speed of analytics without compromising the speed of critical operational business systems.

- **You require always-on analytics.** You may need your analytic application to be always available. Examples include logistics, supply chain, fraud detection, and financial services applications. Full-time availability for a single database can be risky, especially if it doesn't have native failover capabilities. Instead of letting a database become the single point of failure, a distributed data cache provides a more reliable alternative. In this environment when one node goes down, others immediately take over without any interruption in service.

- **You need analysis of big data.** For big data analysis, you may not want to analyze the entire data set where it is stored. One example is analyzing data stored in Hadoop, which while extremely powerful, is subject to high-query latency, making it less-than-ideal for real-time analytics. Instead you want to load a slice of your big data set into memory for speed of thought analysis and visualization. Discover patterns using data cached in memory, then, connect directly to Hadoop for scheduled detail reports and dashboards.

Would You Still Need A Database?

As much as caching data in memory helps with many analytical scenarios, only having in-memory architecture is limiting. You will still need your database.

Here are three reasons why:

- **Big data memory requirements are costly.** Some analytics solutions require you to load all of your data into memory, forcing you to invest in very expensive hardware, or more likely subjecting your analytics solution to scalability constraints. The ideal solution allows you to choose the optimal trade-off between storing data in memory or in a database and allows you to accelerate performance by adding more memory to the system without being subject to the memory size constraints of traditional proprietary solutions.

- **Databases are more powerful when it comes to complex calculations.** With an in-memory only solution, complex calculations on large data sets can easily result in a "out of memory" error. To resolve this you are either forced to get a larger memory capacity, slim down your data sets, or modify your calculations (and as a result spend hours remodeling your data sets).

- **For up-to-the-minute information, you still need your data closest to its source.** If things are changing so fast that you need to see them in real time, you need a live connection to your data. For example, some operational analysis applications like those used by financial services organizations need competitive, real-time, or near-real-time data. Your operational dashboards can be hooked up directly to live data so you know when you are facing peak demand or under-utilization. An all in-memory solution would not provide the latest, freshest data.

Real-time Analytics and the CAP Theorem

Big data refers to the volume, velocity, and variety of highly structured, semi-structured and loosely structured data that is in motion (streaming) and at rest (stored). Most approaches to big data analytics are focused on batch processing of data, in essence big data at rest. This means that analytic results such as trends and patterns only consider what has happened in the past and not what is happening in the present.

What about big data in motion?

If you recall our discussion regarding the CAP theorem in Chapter 5, it stipulates that it is impossible for any distributed computing system to simultaneously address consistency, availability, and partition tolerance; you can at best achieve two out of three. A system with high partition tolerance and availability (like Cassandra) will sacrifice some consistency in order to do it. Similarly, for real-time analytics solutions, there is a variant of the CAP theorem, called SCV (Figure 8-1).

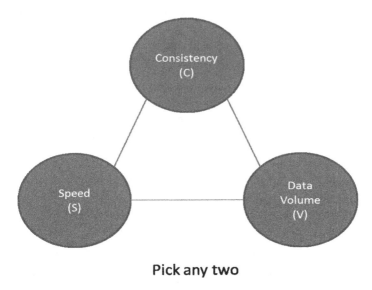

Pick any two

Figure 8-1. SCV for real-time analytics systems

Speed: This is all about fast response times and how quickly you can return an appropriate analytic result from the time it was first observed. In essence, a real-time system will have an updated analytic result within a relatively short time of an observed event, whereas a non-real-time system might take hours or even days to process all of the observations into an analytic result.

Consistency: This is all about confidence level on the accuracy of the response, how accurate or precise (two different things) the analytic outcome is. A totally consistent result accounts for 100 percent of observed data accounted for with complete accuracy and some degree of precision. A less consistent system might use statistical sampling or approximations to produce a reasonably precise but less accurate result.

Data Volume. This is all about the coverage or reach of the analytical result; in other words, this refers to the total amount of observed events and data that need to be analyzed. The problem starts at the point when data starts to exceed the bounds of what can fit into memory. Massive or rapidly growing data sets have to be analyzed by distributed systems.

If your working data set is never going to grow beyond 40 to 50 GB over the course of its lifetime, then you can use an RDBMS or a specialized analytic appliance solutions and have 100 percent consistent analytic results delivered to you in real-time, because your entire working data set can fit into memory on a single machine and doesn't need to be distributed.

However, if you're building an application with a rapidly growing data set and unpredictable burst loads, you're going to need a system that sacrifices some speed or consistency in order to be distributed so it can handle the large volume of raw data.

Batch-oriented analysis is important for certain type of business needs; where you want to do detailed analysis of data, it's more important for the data to be comprehensive (large) and accounted consistently. Whereas in case of real-time analytics you need split second responses, consistency could be compromised with an (+/-) error percentage or confidence percentage.

The one property you should not intentionally sacrifice is data volume, as data is a business asset. Data has inherent value. You want to design your analytic systems to consume and retain as much of it as possible.

Think about this trade-off carefully before you go about building a real-time analytics system.

How Does Real-Time Analytics Work?

Typically when we are talking about real-time or near real-time systems, what we mean are architectures that allow you to respond to data as it's received, without necessarily persisting it to a database first (Figure 8-2).

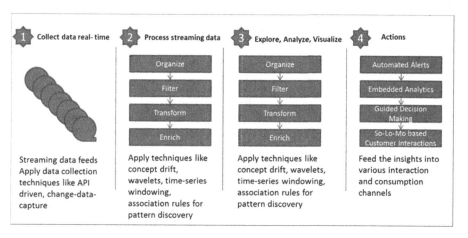

Figure 8-2. *Real-time analytics system processes*

In other words, real-time denotes the ability to process data as it arrives, rather than storing the data and retrieving it at some point in the future. That is the primary significance of the term: "real time" means that you are processing data in the present, not in the future.

Collect real-time data: Big data in motion includes data from sensors, smart grid meters, RSS feeds, computer networks, and social media sites. Real-time data or streaming data can be conceived as a continuous and changing sequence of data that continuously arrive at a system to store or process. The data make up a massive volume (e.g., terabytes), temporally ordered, fast changing, and potentially infinite. For streaming to work, you should have the ability to capture and process streams of data in real time.

Process streaming data: The key to real-time analytics is that we cannot wait until later to do things to our data; we must analyze it instantly. Stream processing (also known as streaming data processing) is the term used for analyzing data instantly as it's collected. Actions that you can perform in real-time include splitting data, merging it, doing calculations, connecting it with outside data sources, forking data to multiple destinations, and more.

Explore, analyze, and visualize data: Now that data has been processed, it is reliably delivered to the databases that power your reports, dashboards, and ad-hoc queries. There are specialized streaming data algorithms and advanced data visualization techniques that you can employ to generate insights.

In real-time systems, scoring is an extremely important activity, and it is triggered by actions (by consumers at a website or by an operational system through an API), and the resulting action or messages are brokered through the consumption channels. During the scoring activity, some real-time systems will use the same hardware that's used for data ingestion, but they will not use the same data. At this phase of the process, the scoring rules are kept separate from the ingested data. Note also that at this phase, the limitations of Hadoop become apparent. Hadoop today is not particularly well suited for real-time scoring, although it can be used for "near real-time" applications such as populating large tables or pre-computing scores.

Data is always changing, so there is a need to refresh the data and refresh the model built on the original data. The existing scripts or programs used to run the data and build the models can be re-used to refresh the models. Simple exploratory data analysis is also recommended, along with periodic (weekly, daily, or hourly) model refreshes.

Refreshing the model based on re-ingesting the data and re-running the scripts will only work for a limited time, since the underlying data and even the underlying structure of the data will eventually change so much that the model will no longer be valid. Important variables can become non-significant, non-significant variables can become important, and new data sources are continuously emerging. If the model accuracy measure begins drifting, you have to go back and reexamine the data. If necessary, go back and rebuild the model from scratch.

Actions: Once you start spotting patterns and anomalies in the streaming data, you need to channel these insights to appropriate consumption channels. This is the layer that most people see. It's the layer at which business analysts, c-suite executives, and customers interact with the real-time big data analytics system.

Real-time big data analytics is an iterative process involving multiple tools and systems.

The Hadoop and NoSQL Conundrum

In earlier chapters we have discussed at length how Hadoop framework helps in analyzing massive sets of data by distributing the computation load across many processes and machines. Hadoop embraces a map-reduce framework, which means analytics are performed as batch processes. Depending on the quantity of data and the complexity of the computation, running a set of Hadoop jobs could take anywhere from a few minutes to many days. Batch processing tool sets like Hadoop are great for doing one-off reports, a recurring schedule of periodic runs, or setting up dedicated data exploration environments. However, waiting hours for the analysis you need means you aren't able to get real-time answers from your data.

Hadoop analysis ends up being a rearview mirror instead of a pulse on the moment.

NoSQL databases are good at enabling fast queries against many terabytes of data, but they have limitations to do SQL-like joins: the ability to combine data from one database table with data from another table. The work around is to de-normalize your datasets. For example, if you are asking a question such as "Find all Twitter posts that contain the phrase "IPL" from all authors based in London, England." In a traditional relational database like SQL, a table of "posts" would join against a table of "authors" using a shared key like an author's ID number. In NoSQL databases, you will de-normalize the data set by inserting a copy of the author into each row of their posts. Rather than joining the posts table with the authors table during the query, all the authors' data is already contained within the posts table before the query.

The question then becomes when should the de-normalization of your NoSQL database occur?

One option is to use Hadoop to append other data sets to the de-normalized data from normalized tables before running these kinds of queries. This approach is fine for batch processing; you still cannot perform complex queries of real-time data. What if we could write de-normalized data on the fly: taking our example of Twitter posts into consideration, write each incoming Twitter post into a row in the posts table, and augment that row with information about the author in real time. This would keep all data de-normalized at all times, always ready for downstream applications to run complex queries and generate the rich, real-time business insights. Real-time analytics and stream processing make this possible.

In the sections below, we will discuss several approaches to process real-time insights.

Using an In-Memory Data Grid for Near Real-Time Data Analysis

Over the last several years, in-memory data grids have proven their value in storing fast-changing application data and scaling application performance. More recently, in-memory data grids have integrated map-reduce analytics into the grid to achieve powerful, easy-to-use analysis and to enable near-real-time decision making.

Data motion to and from a distributed file system increases both access latency and I/O overhead, significantly lengthening the execution time for analysis. In contrast, in-memory data grids perform analytics in place on memory-based data, avoiding data motion and driving down the time required to complete a map-reduce analysis. This enables in-memory data grids to analyze data significantly faster than Hadoop or other file-based analytics platforms, thereby delivering results with minimum latency.

Let's consider the stock trading application example in financial services. The stock trading application receives a market feed of stock price changes occurring during the trading day. This application then applies various analyses to develop a trading strategy to place new trades based on history of price changes for individual stocks and changing risk profiles. In order to develop a recommendation, the stock trading application needs to store a large set of stock histories. Every few seconds the application needs to perform map-reduce analytics across either all or a selected set of stock symbols (such as a market sector), comparing potential returns, evaluating risk profiles, and optimizing the overall trading strategy. This ability to scan a large, fast-changing data set in real time gives the trading analyst an important new tool for detecting changing market conditions and optimizing the selection of trades to place.

Now let's discuss how an in-memory data grid actually works. First of all, it is important to understand that an in-memory data grid is not the same as an in-memory database. Typical examples of in-memory databases are Oracle TimesTen, *SAS* In-*Database* and SAP* HANA. In-memory databases are full database products that simply reside in memory. As a result of being a full-blown database, they also carry the weight and overhead of database management features. In-memory data grid is different: no tables, indexes, triggers, stored procedures, process managers etc., just plain storage.

The data model used in the in-memory data grid is key-value pairs. Unlike traditional systems where keys and values are often limited to byte arrays or strings, with in-memory data grids you can use any domain object as either value or key. Most in-memory data grids are written in Java, thus they have the ability to support a wide variety of data types ranging from simple data types such as a string or number, to complex objects. In-memory data grid has the ability to interface with the distributed data store as with a simple hash map. Being able to work with domain objects directly is one of the main differences between in-memory data grids and in-memory databases. With the in-memory databases, users still need to perform object-to-relational mapping, which typically adds significant performance overhead.

Data consistency is one of the main differences between in-memory data grids and NoSQL databases. NoSQL databases are usually designed on top of the eventual consistency approach where data is allowed to be inconsistent for a period of time as long as it will become consistent eventually. In-memory data grids are positioned as complementary with distributed caching and can be effectively leveraged to support following core patterns:

- **Transactional "write through."** Changes are synchronously updated in databases. The update to the in-memory data grid is successful if and only if the update is also in the database.

- **Asynchronous "write behind."** Queue the updates across the in-memory data grid cluster and transfer the changes in batches to the backend repository. The queuing can be configured to be in-memory replicated for high availability.

Figure 8-3 shows an in-memory data grid with a key set of *{k1, k2, k3}* where each key belongs to a different node.

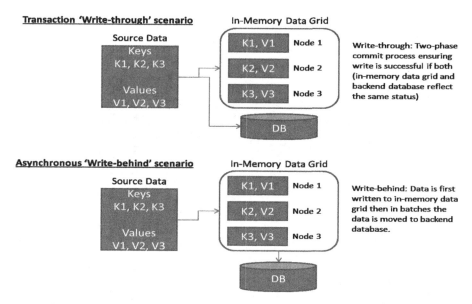

Figure 8-3. *In-memory data grid*

Map Reduce and Real-Time Processing

Hadoop's map-reduce model is very good in processing large amount of data in parallel (Figure 8-4). It provides a general partitioning mechanism (based on the key of the data) to distribute aggregation workload across different machines. Basically, map-reduce algorithm design is all about how to select the right key for the record at different stages of processing.

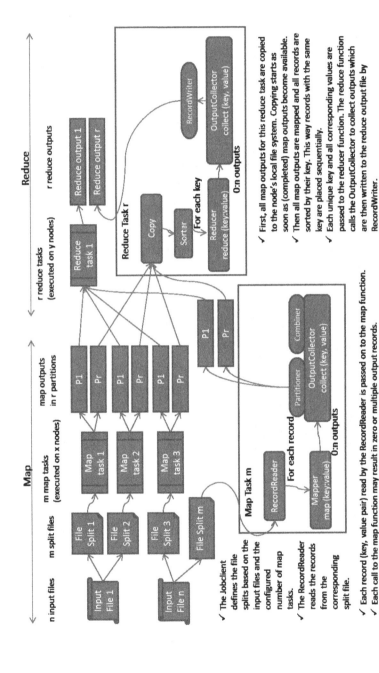

Figure 8-4. Map-reduce processing

However, "time dimension" has a very different characteristic compared to other dimensional attributes of data, especially when real-time data processing is concerned. It presents a different set of challenges to the batch oriented, map-reduce model.

- Real-time processing demands a very low latency of response, which means there isn't too much data accumulated at the "time" dimension for processing.

- Data collected from multiple sources may not have all arrived at the point of aggregation.

- In the standard model of map-reduce, the reduce phase cannot start until the map phase is completed. And all the intermediate data persists in the disk before download to the reducer. All these added to significant latency of the processing.

Although Hadoop map-reduce is designed for batch-oriented work load, certain applications, such as fraud detection, ad display, network monitoring requires real-time response for processing large amount of data, application designers have started to looked at various way of tweaking Hadoop to fit in the more real-time processing environment. We will discussfew techniques to perform low-latency parallel processing based on the map-reduce model.

There is another aspect of low latency - How current the analyzed data is; in the case of HDFS it is as current as the last snapshot copied into it. The need for snapshotting comes from the fact that most businesses are still running on traditional RDBMS systems (with NoSQL gaining momentum recently), and data has to be at some point migrated into HDFS in order to be processed. Such snapshotting is currently part of most Hadoop deployments and it usually happens once or twice a day.

By putting an in-memory data grid in front of HDFS we can store recent or more relevant data in memory, which allows for instant access and fast queries on it (Figure 8-5). When the data is properly partitioned, you can treat your whole in-memory data grid as one huge memory space: you can literally cache terabytes of data in memory. But even in this case, the memory space is still limited and when the data becomes less relevant, or simply old, it should still be offloaded onto RDBMS, HDFS, or any other storage. With this architecture, businesses can now do processing of both, current and historic data.

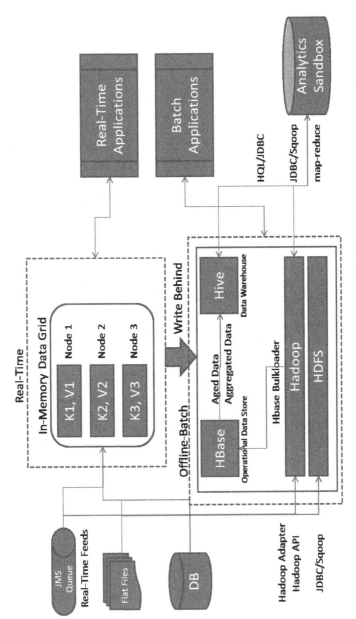

Figure 8-5. *In-memory data grid and Hadoop ecosystem conceptual architecture*

If we do not want to use in-memory data grid and still achieve real-time analysis, there is an alternative approach as well. In a typical Hadoop implementation, you will notice map-reduce jobs are executed in a scheduled manner to run against the data stored in HDFS. HFlame enhances Hadoop core with real-time streaming analysis capability. In traditional Hadoop, a map-reduce job processes only the current snapshot of available data and ends right after it finished processing the snapshot. Processing of any new contents requires scheduling of another map-reduce job. With HFlame enhanced Hadoop, map-reduce jobs can optionally be configured to run in continuous mode. Which essentially means that map-reduce job doesn't end even if there are no more new contents available. As soon as new data is pushed in HDFS, continuously running map-reduce jobs are notified, which immediately passes the new contents through map-reduce process and extract insights.

HFlame supports following behavior:

1. HFlame runs on top of customer's Hadoop installation. HFlame is an incremental add on to existing Hadoop clusters.

2. No new API. Completely driven by configuration.

3. HFlame's real time map-reduce jobs are completely fault tolerant. In the event of any failure, failed components are automatically scheduled on other available Hadoop nodes.

4. HFlame guarantees no data loss. If any component of map-reduce job or Hadoop infrastructure fails in the middle, automatic job/component's recovery procedure will take of care starting the data processing from exactly the same place where it failed.

5. Allows building a complex mesh of real time map-reduce jobs to support data analysis requirements that cannot be described in single map-reduce process.

6. Supports data analysis frameworks like PIG, HIVE.

7. Real time map-reduce jobs can optionally be run in batch mode, i.e., reduce tasks, accumulate data for a certain amount of time, and then produce the aggregated results.

Figure 8-6 explains the flow of real-time map-reduce job.

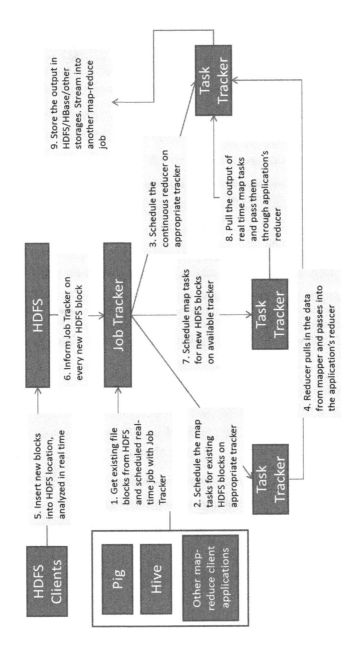

Figure 8-6. Real-time map-reduce job flow

The HFlame compelling argument is the common data analysis framework for both offline and real-time massively parallel data analysis, which essentially means no new storage, no new data processing semantics, and leveraging existing high-level abstraction languages like Pig and Hive. For Hadoop users, real-time streaming analysis with HFlame requires absolutely zero investment into new infrastructure and no new API/tools to learn.

Use Case: Real-Time Analysis of Machine Generated Data (Log Processing)

Machine data (or data exhaust) is produced all the time by nearly every software application and electronic device. The applications, servers, network devices, sensors, browsers, desktop and laptop computers, mobile devices, and various other systems deployed to support operations are continuously generating information relating to their status and activities.

Machine data is generated by both machine-to-machine (M2M) as well as human-to-machine (H2M) interactions. Machine data in is generated in a multitude of formats and structures, as each software application or hardware device records and creates machine data associated with their specific use. Machine data also varies among vendors and even within the same vendor across product types, families, and models. The figure below illustrates the type of machine data created and the business and IT insights that can be derived when a single web visitor makes a purchase in a typical e-commerce environment shown in Figure 8-7.

Event	Machine Generated Data	Business/IT Insights
Customer adds products to shopping cart	**Product Action User Session User Browser Information** IP:...xxxxxxxxx..ababaaba Cart:"Big Data Imperatives" ProdID:123456xxxx.xxxx.xxxx xxxxxx...xxxxxx.xxxxxxx Time:xxxxx.xxxx.xxxx.	Product performance Shopping cart activity Website performance
Customer completes transaction through credit card	**Social ID User email User City Product** IP:..xxxxxxxxx..ababaaba TwitterID:xxxxx.xxxx.xxxx Email:xxxxxx...xxxxxx.xxxxxxx Address:xxxx.xxxx ProdID:123456xxxx.xxxx.xxxx Time:xxxxx.xxxx.xxxx.	Customer behavior Inventory updates Fraud heuristic analysis Real-time product sales tracking
Webserver attempts to write to database	**Error Error Details** IP:..xxxxxxxxx..ababaaba AppErrorID:xxxxx.xxxx.xxxx AppErrorMsg:xxxxxx...xxxxxx.xxxxxxx Time:xxxxx.xxxx.xxxx.	Nature of application error Source of error

Figure 8-7. Machine-generated data and business impacts

Figure 8-7 is an example of the type and amount of valuable information generated by a single website visitor that is recorded. A typical e-commerce site serving thousands of users a day will generate gigabytes of machine data that can be used to provide significant insights into the IT infrastructure and business operations.

How do we process machine-generated data?

Let's consider the scenario of a retailer. The retailer is using around forty applications hosted on a multitude of servers (200+) in different data centers to manage their business processes. A single business process involves several business applications, workflows, and associated data. Each application creates log messages that are stored as text files in the local file system. There are multiple web applications where customers browse products and offerings and select their preferences. Once an order is placed, the request is sent to a central order-processing application. This application performs the following steps in order: checks for availability in the inventory management application, performs the payment in the credit-card-processing application and initiate the shipment. Each of the involved applications runs on different servers and produces log files.

The log messages usually consist of some fixed fields, like for example a timestamp, a logging level, or the name of the logging component or application in addition to key information in unformatted plain text. There are different possibilities for storing these messages. In a relational database, you might just record the whole of the message in a CLOB field, or you might try to store the message into a pre-defined schema. While CLOB will be an easy way to store the data, retrieving information from CLOB has additional challenges, as you will have to develop a full text search index to be able to find a message in the CLOB field. The approach with the pre-defined schema also has several disadvantages, leading to problems when adding a new application that creates log messages with a different format. Messages might also be truncated when they do not fit into the schema.

Figure 8-8 shows an example of a log message. Each message consists of a couple of fields like timestamp, class, method, and log level, followed by the log message itself consisting of unstructured text. Different applications can have different sets of logging fields.

Figure 8-8. Log message example

Log data is generated continuously, and so they need to be captured the moment they are generated; modifications of log data might happen but are very rare. In an environment where hundreds or thousands of log messages are stored every second this task can be difficult to accomplish. It will quickly become necessary to partition the database schema to distribute the load across multiple hard disks or even servers. Partitioning log data does not work well with a time-based partitioning scheme, because most of the data is inserted in the partition that contains data for the current date. To distribute the load, it is necessary to use another schema, for example by partitioning via the applications. But here we have to deal with the problem that different applications produce different amounts of log data. Finding a balanced partitioning scheme is a challenge and partitioning schemes may change over time.

To develop a real-time log management system, our first task is to find a file format that allows fast and direct access to a single log message while storing hundreds of thousands log message in a single file. Hadoop/HDFS could be the solution we are looking for (Figure 8-9).

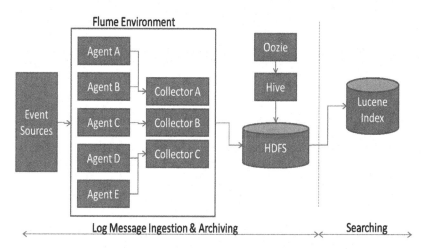

Figure 8-9. *Log processing, Hadoop, and search conceptual architecture*

Hadoop provides two file formats for grouping multiple entries in a single file:

- **SequenceFile:** A flat file which stores binary key-value pairs. The output of map-reduce tasks is usually written into a SequenceFile.

- **MapFile:** Consists of two SequenceFiles. The data file is identical to the SequenceFile and contains the data stored as binary key-value pairs. The second file is an index file, which contains a key-value map with seek positions inside the data file to quickly access the data.

The SequenceFile format seems to be well suited for storing log messages and processing them with map-reduce jobs; but the direct access to specific log messages is very slow. The API to-read data from a SequenceFile is iterator based, so that it is necessary to jump from entry to entry until the target entry is reached. One of the most important use cases is searching for log messages in real time, as slow random access performance is a showstopper.

In contrast to SequenceFiles, MapFiles uses two files; the *index file* stores seek positions for every n-th key in the datafile. The *data file* stores data as binary key-value pairs. However, using MapFiles comes with a disadvantage, which is that any instance of a random access needs to read from two separate files. This process seems to be slow, but the indexes that store the seek positions for log entries are small enough to be cached *in memory* (Figure 8-10). Once the seek position is identified; only relevant portions of the data file are read.

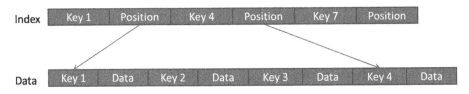

Figure 8-10. *Index and data mapping*

Since MapFiles and SequenceFiles use binary key-value pairs we need a data format to store log messages in these files. In order to be able to search efficiently for log messages, you need to store data fields as separate entities. Google protocol buffers provide excellent functionalities to transfer and store log messages. Protocol buffers are encoded structured data.

Below listed are few most important reasons for choosing the Google protocol buffer format:

- **Speed:** Deserialization speed is one of the most important factors when evaluating file formats. Especially map-reduce jobs that crunch through the whole data set stored in the HDFS rely on fast object deserialization. Protocol buffers make up one of the fastest frameworks. Object deserialization with Protocol buffers is sixteen times faster than with pure Java serialization.

- **Size:** Ability to store billions of serialized objects is another key factor. Protocol buffers produce serialized objects that are around four times smaller than those produced by the standard Java serialization.

- **Migrations:** One unique feature of protocol buffers is the ability to change the file format without losing backward compatibility. It is possible to add or remove fields from an object without breaking working implementations. This is a very important feature when serializing objects for long-time storage.

- **Platform- and Language-independent:** Protocol buffer objects may be accessed from multiple languages on any operating system. This feature allows us to use protocol buffers as the sole data format throughout the whole log processing chain.

Log messages consist of textual data which can be compressed very efficiently. MapFiles and SequenceFiles both offer a transparent compression mechanism, it is possible to compress each log entry individually or use a block-level compression where multiple entries are compressed together. Let us assume our log messages have an average size of 500 characters. Using block compression, it is possible to drop the size down to around 20 percent of the original message size. Hadoop's default setting uses a block size of 1 MB uncompressed data, which in our case means around 2,000 log messages are compressed together in a single block.

■ **Note** There is a downside to using block compression. With block compression each block has to be read completely and decompressed before a single entry may be accessed. This is not really a problem on its own, but combined with the seek positions stored in the index file it starts to be a problem.

Index files store the seek position of each n-th entry in the datafile. This means that in order to save memory for example, only each 16th entry will be written in the index file. The seek position is the position where the data may be found in the compressed data file. If we have 2,000 log entries stored in a single block and the position of each 16th entry is written to the index file, we have 125 entries with identical seek positions (all 2,000 log entries start at the same block and have identical seek positions), which is a waste of memory. This is known as the "seek position grouping" issue.

On the other hand when 2,000 entries are found at the same seek position, we need to iterate over the entries stored in a block until we reach our requested entry. This seriously impacts random access performance.

■ **Note** It is absolutely necessary to tweak the block compression size and the skip rate in the index file to find the optimal compromise between compression factor and the number of entries per block. The skip rate should be chosen so that each block has only one entry in the index file.

How to perform near real-time searches on up to 36.6 billion log messages?

A typical search requirement is that 95 percent of all search queries *should display the results in less than 10 seconds.* These requirements are difficult to implement with a pure Hadoop solution. In Hadoop we can use map-reduce jobs to retrieve data. A map-reduce job that has to read all the data may run for about hours. The only way to be able to search in real time is to build a search index on the stored data. *Lucene,* a specialized search framework seems to be a very good partner for Hadoop. It is implemented in Java, which means a very good integration with Java-based application. Lucene is also highly scalable and has a powerful query syntax.

Now let's look at how our solution works!

Lucene is able to distinguish multiple indexed fields in a single document. Log data can be split up in distinct fields like the timestamp of the message, the log level, the message itself, etc. A Lucene index consists of documents. Each document has a number of fields. The contents of a field can consist of one or more terms. The number of unique terms is on criteria for the memory requirements of an index.

Each document needs to have a primary key field, which specifies how the document can be retrieved. The primary key field contains the full path inside the HDFS to the MapFile, which contains the log message, followed by the index of the log message inside this MapFile (Figure 8-11). This enables us to directly access the referenced log message.

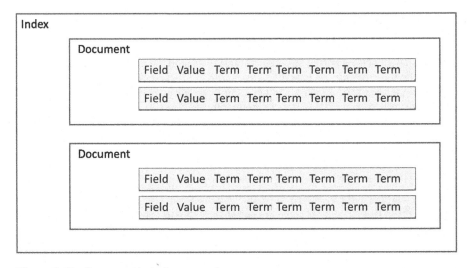

Figure 8-11. *Document indexing example*

Lucene can build up a full text index of rather large files. However, one must pay attention to the memory requirements. The memory requirements depend heavily on the number of indexed fields, the type of the indexed fields, and whether the contents of a field have to be stored in Lucene or not.

By analyzing typical search queries you can identify fields that need to be indexed in order to run 95 percent of all search queries in real time. At a broad level you only need six fields in the Lucene index:

- timestamp of the log message

- numeric ID of the application that created this log message

- numeric log level

- name of application server the application is running on

- host name of the server the application is running on

- path in the Hadoop file system where the complete log message can be read

The next step is to optimize the memory requirements for each field. The timestamp field can become a bit tricky to manage especially if you are capturing to the level of milliseconds. This will lead to a huge number of unique values inside the timestamp field and will impact memory requirements. On the other hand, in the search queries timestamps are only specified up to an accuracy of minutes. You need the higher accuracy to sort the search results.

A solution could be to split the timestamp field into two separate fields in the Lucene index. One field stores the timestamp rounded up to minutes and is indexed. The other field stores the timestamp with full accuracy and is only stored in Lucene, not indexed. With this solution you can reduce the number of unique terms that Lucene needs to handle and therefore greatly reduce the impact on memory requirements. Another benefit of this approach is increased performance when searching for date ranges. The downside is that you need to sort the result set yourself using the detailed timestamp field after getting the search results from Lucene.

In order to have high availability, we can split up the Lucene index into smaller parts which can be served on each datanode. We can further allocate 6 GB of heap memory on each data node to Lucene so that each data node is able to run the index for up to 1 billion documents. *Solr* is a search platform based on Lucene. It provides a web-based interface to access the index. This means we can use a simple HTTP/REST request to index documents, perform queries, and even move an index from one data node to another. Each data node is running a single Solr server that can host multiple Lucene indexes. New log messages are indexed into different shards, so that each index has approximately the same number of documents. This approach balances the load on each shard and enables scalability. When a new data node is integrated into the cluster, the index shard on this data node will be primarily used for indexing new documents.

For performance reasons, the index data files are stored on the local file system of each data node (Figure 8-12). Each time an index has been modified, it will be backed up into the Hadoop file system. Now we are able to quickly redeploy this index onto another data node, in case the data node which originally hosted this index has failed.

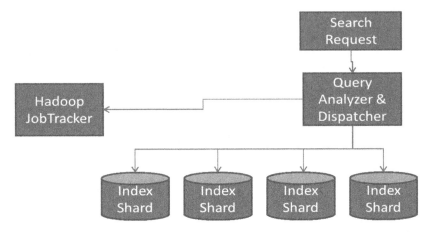

Figure 8-12. *Index sharding example*

An incoming search query is analyzed for the queried fields. If all fields are indexed, parallel search queries will be sent to all index shards. The responses will be collected, sorted, and then returned to the user. If the search fields are not indexed, a new map-reduce search job will be created and submitted to the Hadoop job tracker.

When performing a query, all index shards are queried in parallel. This ensures fast response times. When a user formulates a query, it is first analyzed if this query can be run against the Lucene indexes. This is not the case if the user specifies search fields which are not indexed. In that case the query will be run as a map-reduce job. If the query can be run against the Lucene indexes, it will be forwarded to all data nodes in parallel. The results of these subqueries are collected and sorted. Then the log messages are read from the HDFS using the primary keys inside the Lucene index results.

If a query to a single shard fails, the search results may be incomplete, but the queries to the other shards are not affected. This greatly enhances the availability of the system.

Building a Recommendation System

With the number of options available to the users is ever increasing, the attention span of customers is getting lower and lower. Customers are used to seeing their best choices right in front of them. In such a scenario, we see recommendations powering more and more features of the products and driving user interaction. Hence companies are looking for more ways to minutely target customers at the right time. Some of the examples of recommendation systems include product recommendations, merchant recommendations, content recommendations, social recommendations, query recommendation, display and search ads (Figure 8-13).

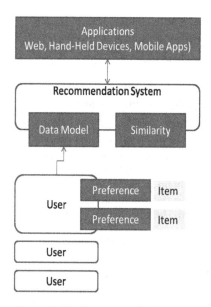

Figure 8-13. *Recommendation system conceptual architecture*

This brings big data into the picture. Succeeding with data and building new markets, or changing the existing markets is the game being played in many high-stakes scenarios. Some companies have found a way to build their big data recommendation/machine-learning platform, giving them the edge in bringing better and better products even faster to the market. The more data we give to our algorithms, the better-targeted results we get. A recommendation platform using Hadoop would have the following components: ETL, feature generation, feature selection, recommendation algorithms, A/B testing, serving, tracking, and reporting.

In the sections below, we go over use cases and details of solving them in the Hadoop ecosystem. We will also specifically cover a set of machine-learning algorithms for solving the various recommendation use cases. While Mahout fits well with Hadoop map-reduce framework, there are also elegant ways of plugging in other non-distributed systems/algorithms into Hadoop.

Let's first review some basic concepts related to recommendation system.

In a classical model of recommendation system, there are "users" and "items." A "user" has associated metadata (or content) such as age, gender, race, and other demographic information. "Items" also has its metadata, such as text description, price, weight, etc.

On top of that, there are interactions (or transactions) between the user and items, such as user A downloading/purchasing item X or user A giving a rating 5 to a product Y. In a real-world scenario, you will find many-to-many relationships between users and products.

Now given all the metadata of user and item, as well as their interaction over time, can we answer the following questions?

- What is the probability that user X will purchase item Y?

- What rating will user X give to item Y?

- What is the top k unseen items that should be recommended to user X?

Content-based Approach: In this approach, we make use of the metadata to categorize user and item and then match them at the category level. One example is to recommend jobs to candidates; we can do an IR/text search to match the user's resume with the job descriptions. Another example is to recommend an item that is "similar" to the one that the user has purchased. Similarity is measured according to the item's metadata, and so various distance functions can be used. The goal is to find k nearest neighbors of the item we know the user likes.

Collaborative Filtering Approach: In this approach, we look solely at the interactions between user and item and use that information to perform our recommendation. The interaction data can be represented as a matrix.

Notice that each in Table 8-1, the cells represent the interaction between the user and the item. For example, the cell can contain the rating that the user gives to the item (in that case, the cell is a numeric value), or the cell can be just a binary value indicating whether the intersection between a user and an item has happened. (e.g., a "1" if user X has purchased item Y, and "0" otherwise.)

Table 8-1. *User-Item Matrix*

	Item 1	Item 2	Item 3	Item 4	Item 5	Item 6
User 1	1	1	0	0	0	1
User 2	0	0	1	0	1	1
User 3	1	0	1	1	1	0
User 4	0	0	1	1	1	1

The matrix is also extremely sparse, meaning that most of the cells are unfilled. We need to be careful about how we treat these unfilled cells. There are two common ways of treating them:

- Treat these unknown cells as "0". Make them equivalent to a user giving a rating of "0". This may or may not be a good idea depending on your application scenarios.

- Guess what the missing value should be. For example, to guess what user X will rate item A, given we know his rating on item B, we can look at all users (or those who are in the same age group of user X) who have rated both item A and item B then compute an average rating from them. Use the average rating of item A and item B to interpolate user X's rating on item A given his rating on item B.

User-based Collaboration Filter: In this model, we do the following:

1. Find a group of users that is "similar" to user X

2. Find all movies liked by this group that hasn't been seen by user X

3. Rank these movies and recommend to user X

This introduces the concept of user-to-user similarity, which is basically the similarity between two row vectors of the user/item matrix. To compute the k nearest neighbor of a particular user, a naive implementation is to compute the "similarity" for all other users and pick the top k.

Different similarity functions can be used. The Jaccard distance function is defined as the number of intersections of movies that both users has seen, divided by the number of unions of movies they have both seen. Pearson similarity first normalizes the user's rating and then computes the cosine distance.

Item-based Collaboration Filter: If we transpose the user/item matrix and do the same thing, we can compute the item to item similarity. In this model, we do the following:

1. Find the set of movies that user X likes (from interaction data)

2. Find a group of movies that are similar to the set of movies that we know user X likes

3. Rank these movies and recommend them to user X

It turns out that computing the item-based collaboration filter has more benefit than computing user-to-user similarity for the following reasons:

- The number of items is typically smaller than the number of users

- User's tastes will change over time, and so the similarity matrix needs to be updated more frequently. Item-to-item similarity tends to be more stable and requires fewer updates.

Singular Value Decomposition: If we look back at the matrix shown in Table 8-1, we can see the matrix can be viewed as multiplications of items from the item space with users from the user space. In other words, if we view each of the existing items as an axis in the user space, then multiplying a new item within the matrix results in a vector similar to user. We can then compute a dot product with a new item with the same set from the user space to determine its similarity. If we keep the item space in the matrix the same and map a new user to the item space, we follow the same approach to compute a dot product that will result in a vector similar to item space.

Association Rule Based: In this model, we use the market/basket association rule algorithm to discover rules like ... {item1, item2} => {item3, item4, item5}.We represent each user as a basket and each viewing as an item (notice that we ignore the rating and use a binary value). After that, we use the association rule mining algorithm to determine what the frequently occurring items in the overall data set are and what association rules can be defined for these frequently occurring items. Then for each user, we match the user's previous items viewed to the set of rules to determine what other movies we should recommend.

Referring to Figure 8-2, the recommender retrieves items and users from the data model. The data model provides methods that count the total number of users, total number of items, number of users that prefer a certain item, etc. Similarity functions use these numbers to compute a similarity value for pairs of items or users. We discussed several algorithms you can choose from to build a recommender. However, the *Mahout Tanimoto Coefficient Similarity*, is a relatively straightforward similarity algorithm that is widely used in recommendation systems for discovering similarities. Let's illustrate the algorithm in the context of a webshop. Suppose there are three customers, A, B, and C, and five products, numbered one up to five. Say each customer has bought a few products. For this algorithm it does not matter how many products are purchased, only which products are purchased by which customer.

Table 8-2. *Customer-Product Matrix*

	Customer A	Customer B	Customer C
Product 1	ü		ü
Product 2	ü	ü	
Product 3		ü	
Product 4	ü		
Product 5	ü	ü	ü

Intuitively you may see that the similarity between two products can be expressed by some ratio of purchases of customers. Simply put, the Tanimoto coefficient uses the ratio of the intersecting set to the union set as the measure of similarity. Represented as a mathematical equation:

$$T(a,b) = \frac{N_c}{N_a + N_b - N_c}$$

where
Nc = Number of customers that purchased p1 and p2,
Na = Number of customers that purchased p1, and
Nb = Number of customers that purchased p2

This means that if many customers have bought both the products, the numerator will be higher and so will be the similarity value. Alternatively, if many people have bought p1 and many have bought p2, but very few people bought both, then p1 and p2 are probably dissimilar. Table 8-3 shows the calculated Tanimoto coefficients for each product pair.

Table 8-3. Tanimoto Coefficients for Each Product Pair

	Product 1	Product 2	Product 3	Product 4	Product 5
Product 1	1	1/3	0	½	2/3
Product 2	1/3	1	1/2	½	2/3
Product 3	0	1/2	1	0	1/3
Product 4	1/2	1/2	0	1	1/3
Product 5	2/3	2/3	1/3	1/3	1

Figure 8-14 is a very high-level view of the architecture diagram.

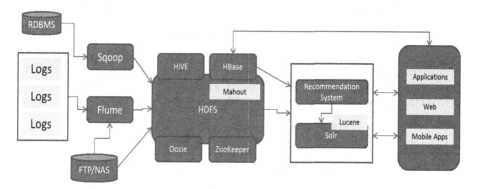

Figure 8-14. Conceptual architecture for real-time log processing and recommendation system

Mahout has quite an extensive set of algorithms that can be run on Hadoop. These include clustering, collaborative filtering, and classification. Hadoop provides the ideal platform for the training and testing of the models. Automating this process with Hadoop brings huge savings in development and operational costs. Tracking and reporting of the performance of the various models significantly helps in knowing how well the system is operating.

End Points

This chapter discussed several application scenarios of big data analytics platforms: in-memory solutions, real-time analytics solutions, and recommendation systems. We discussed a few new technology components like in-memory data grid, Flume, Mahout, Lucene, Solr, etc., which are complimentary to Hadoop/map-reduce.

Big data analytics as an area are fast evolving; in the future, there will be many more technologies that can co-exist and compliment the Hadoop/map-reduce framework. The key is to understand the use case you require. The requirements of use case will dictate what kind of technology components you would like to explore and how effective they are.

References

When Should Analytics be in memory?: www.dbms2.com

The coming in-memory database tipping point: SQL Server Blog – David Campbell

In-memory Analytical Systems: Perspective, Trade-offs and Implementation: TibcoSpotfire Whitepaper

In-Memory Analytics Strategies for Real-Time CRM: Whitepaper by Booz & Company

In-Memory or Not In-Memory - What Should You Expect from Your Business Analytics Application? www.pentaho.com

Plattner, H. and A. Zeier. 2011. *In-Memory Data Management: An Inflection Point for Enterprise Applications*. Heidelberg, Dordrecht, London, New York: Springer.

http://vipuljain99.wordpress.com/2012/10/30/hadoop-analytics-is-not-real-time-a-reality-or-myth-3/

http://vipuljainblogs.blogspot.in/2012/11/storm-s4-or-hflame-real-time-streaming.html

http://blog.mgm-tp.com/2010/06/hadoop-log-management/

http://java.dzone.com/articles/recommendation-engine-models

CHAPTER 9

▓ ▓ ▓

Data Scientist

The realm of big data analytics is vastly different from transaction processing applications and BI applications; here, one discovers and answers questions in area where we don't know what we don't know. The skills required to do these kinds of activities are unique and certainly multi-faceted.

On a general level we can define data as having three important characteristics: composition, context, and condition. Composition refers to the structure of the data: what is the source, what is the granularity, what are the data types, what is the nature of the data (mostly static data or real time streaming data), etc. Context refers to how it was generated, what events are associated with the data, how sensitive the data is, etc. Condition refers to the state of the data and whether it can be used as-is for analysis or it needs further cleansing and enrichment.

Let's apply these characteristics to small data and big data. Small data consists of mostly known data sources that are not expected to undergo changes in composition and context over a given period of time. Since there is a fair amount of certainty regarding small data, we use it solve specific problems through straightforward applications (transaction processing applications, BI reporting, etc.). In essence, small data is limited to answering questions about what we know we don't know. Big data, on the other hand, represents multiple and unknown data sets. These data sets continuously exhibit changes in composition, context, and condition. Thus big data signifies the complexity: we don't know what we don't know!

The biggest problem on hand is how to derive value from big data and finding a way to measure the amount of knowledge contained in data.

A measure of the amount of knowledge contained in data can possibly be defined as the number of insights one can generate by exploiting all the possible range of values (combination and/or permutations) contained within the attributes of the data set. The relative knowledge contained within two variables (A and B), for example, can be assessed by looking at A alone, then B alone, and then A and B, for a total of three scenarios. Three variables (A, B, and C) gives use a knowledge state space of seven. Four subjects results in 15. And so on.

Data analysis is not a new skill. For decades, quantitative research drove the analytics continuum. This was the realm of mathematicians, statisticians, and pure quantitative scientists. The development and enhancement of sophisticated algorithms to solve real-world problems was mostly the purview of academia and research institutions. These were the people with advanced academic degrees who spent years in doing research to further enhance earlier established algorithms like Hidden Markov Support Vector Machines, Linear Dynamical Systems, Spectral Clustering, Machine Learning algorithms or come up with newer models. Once these developments came out of the labs, commercial organizations and product vendors adopt them to make it usable for enterprises.

However, the richness and vastness of big data posed several challenges, namely too many unknowns about the data itself; hence, instead of predefined ways of analyzing data, discovery types of analysis were needed, thus giving rise to a new occupation called "data scientist."

Data scientists are the practitioners of the analytics models solving business problems. They incorporate advanced analytical approaches using sophisticated analytics and data visualization tools to discover patterns in data. In many cases, these practitioners work with well-established analytics techniques such as logistic regression methods, clustering methods, and classification methods to draw insights from data. These practitioners have deep understanding of the business domain and apply that effectively to analyze data and deliver the outcomes in a business understandable intuitive manner through advanced data visualization tools.

A big data scientist understands how to integrate multiple systems and data sets. They need to be able to link and mash up distinctive data sets to discover new insights. This often requires connecting different types of data sets in different forms, as well as being able to work with potentially incomplete data sources and cleaning data sets to be able to use them.

The big data scientist needs to be able to program, preferably in different programming languages such as Python, R, Java, Ruby, Clojure, Matlab, Pig or SQL. They need to have an understanding of Hadoop, Hive and/or Map-Reduce. In addition they need to be familiar with disciplines such as:

- Natural Language Processing: the interactions between computers and humans

- Machine learning: using computers to improve as well as develop algorithms

- Conceptual modeling: to be able to share and articulate modeling

- Statistical analysis: to understand and work around possible limitations in models

- Predictive modeling: most of the big data problems are about being able to predict future outcomes

- Hypothesis testing: being able to develop hypotheses and test them with careful experiments

The exact background of a big data scientist is of less importance. Great big data scientists can have different backgrounds such as econometrics, physics, biostatistics, computer science, applied mathematics, or engineering. However, to be successful, big data scientists should have at least some of the following capabilities:

- Strong written and verbal communication skills

- Be able to work in a fast-paced multidisciplinary environment, because in a competitive landscape new data keeps flowing in rapidly and the world is constantly changing;

- Have the ability to query databases and perform statistical analysis

- Be able to advise senior management in clear language about the implications of their work for the organization

- Have at least a basic understanding of how business strategy works

- Be able to create examples, prototypes, and demonstrations to help management better understand the work

- Have a good understanding of design and architecture principles;

In short, the big data scientist needs to have an understanding of almost everything. Depending on the industry the big data scientist wants to work in, the need to specialize will be even more important; for example, a capital markets big data specialist requires a different set of skills than a big data scientist working in the retail chain area.

The perfect big data scientist who has all of the above-described skills and capabilities is extremely rare. Perhaps only a handful of big data scientists have all the skills as mentioned above. Therefore, organizations should choose and pick from this list what they deem most important in a big data scientist and what the particular requirements are for the job.

The New Skill: Data Scientist

There are several definitions of a data scientist. We will adopt the following definition: "A data scientist is a person who takes raw materials (in this case data) and uses skill, knowledge, and vision to craft it into something of unique value."

Contextualizing data is at the core of the set activities the data scientist performs. The data scientist creates data and analysis workflows that provide the foundation for discovery, whether delivering an answer to a specific question or creating a new application for business users. They not only understand their organization's business drivers and problems but also know where to find the relevant data (internal and external).

We discussed earlier that the data scientists require a combination of technical and business skills. Because the skill set is so diverse, expecting to find these skills in abundance is futile.

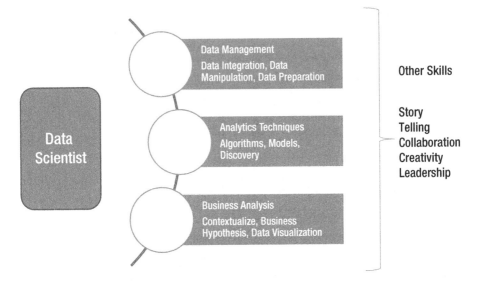

Figure 9-1. *Data scientist skills*

Data Management. At the heart of analytics is data, and so robust data sets are needed for deep analytic efforts. Data can be in disparate locations (internal and external), large in volume, or streaming. Data scientists need to employ several approaches to develop the relevant data sets for analysis. In many cases, data needs to be massaged and prepared to reflect relationships and contexts; these things will not be present in raw transactional data, hence the data scientists need to be good at data integration, data manipulation, and data preparation skills. In one of the earlier chapters we discussed the importance of data quality in preparing data sets for analysis, and so the data scientist needs to have skills to perform profiling, data validations, and cleansing of data.

Analytics Techniques. In the previous chapter we discussed several analytics techniques. Depending on the business problem you are trying to solve and type of data available for you, a broader or narrower set of analytics techniques or algorithms and models will have to be developed. The data scientist needs to be skilled in the various analytics techniques and processes.

Business Analysis. Business context behind the data is the most critical skill a data scientist can possess, because if you do not understand the business attributes of the data, you will not be able to leverage the value of data. In big data scenarios it is easy to get lost in the discovery process when you are dealing with a vast volume or variety of data. The data scientist must have the ability to distinguish "cool facts/analysis" from insights that will matter to the business and to communicate those insights to business executives.

Beyond these three core skills, a data scientist should also possess several other soft skills: storytelling, collaboration, creativity, and leadership.

Data scientists have a difficult job of formulating the right data sets; this means they need to obtain access to the data, work with business users to contextualize the data associating the business meanings behind the data and then explain the findings of their analysis to business stakeholder in a language they understand. In short, the data scientist must have the ability to bring the scenarios to life by using data and visualization techniques: this is nothing but storytelling. They also need to effectively collaborate across several stakeholders within an enterprise (business and technology). Somebody within the enterprise may be holding a vast knowledge of business context behind the data patterns, but the data scientists need to transcend the statistics and mathematics realm and effectively collaborate with these persons.

To solve complex problems, find patterns within volumes of data, and develop intuitive and easily understandable data visualization, the data scientist must be innovative in his/her thinking. The creativity element is very critical; think outside the box, otherwise you end up looking at the data with the same pair of eyes and same thoughts without realizing that the data is actually revealing some interesting aspects. The data scientist should also have enough leadership qualities to emphatically position the findings in front of senior management within the enterprise. Often the data scientist needs to put together a team of data management resources and business analysts to solve a complex problem. In such situations one must have the ability to lead a team and manage the efforts of teams of statisticians, data administrators and integration professionals, and data visualization, reporting, and application integration developers.

The Big Data Workflow

A big data platform can provide a rich data ecosystem by combining data from traditional data warehouses. As far as unstructured data, machine-generated data, and free-form text are concerned, finding answers from this enriched and vast data platform is not a trivial pursuit.

In general, data analysis has many constituent parts. Data must be acquired from myriad sources and cleansed. It must be sorted and joined so that queries can be made against it. It needs to be stored in persistent repositories. Analysts and programmers must then work together in a statistical environment such as R, SAS, or SPSS to query the data.

Then the data must be visualized in some format—a static report, or perhaps in a 2D or 3D visualization tool. The problem is that all of this work with data is not done by a business analyst alone. It is in large measure done by a team of specialists behind the scenes in IT, and every step in this process requires getting someone else involved, who already has a substantial backlog of work.

To the above process, when we add big-data-related unstructured data sources and streaming data, etc., the complexity of managing the activities increases multifold and involves a number of handoffs, resulting in delays based on high demand for specialized data and analytic skills. The person closest to the business user, the data analyst or business analyst, can't do most of the work, and so the time from question to insight involves numerous delays. In fact, it is often the case that decisions are made based on limited information long before the answers come back from the data analysis workflow.

Figure 9-2 is a representation of current practices adopted in a data analysis workflow which can be contrasted with the workflow in a big data setting as shown in Figure 9-3.

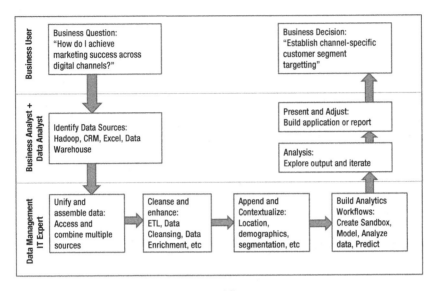

Figure 9-2. *Traditional data analysis workflow*

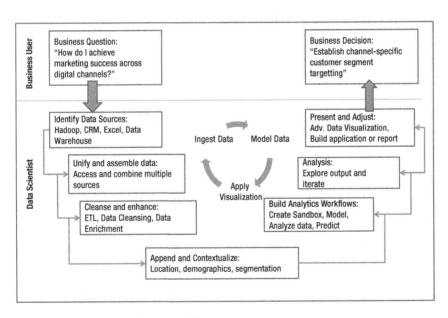

Figure 9-3. *Big data analytics workflow*

Design Principles for Contextualizing Big Data

Contextualizing big data involves several key design principles when it comes to creating solutions that deliver real insight:

- **Ingest and integrate data from anywhere:** Companies are expanding their quest for data into data sources that were previously never considered, and they are going beyond the "system of records" to "system of engagements." The objective should be to develop rich data sets by combining the qualitative structured data with the important context provided by unstructured data.

System of record denotes to data contained within the corporate firewalls, which are of high quality, cleansed and have well defined structure associated with it.

Systems of engagement denotes to data sources and applications that are very much a part of the business eco-system but stay outside the corporate firewalls. The associated data is often unstructured, not well defined and not quality controlled.

- **Discover and seek patterns.** Big data analysis use cases do not follow a predefined path of analysis; they are always led by a train of thought finally leading to the insight generation. While doing this kind of analysis, one does not look for accuracy and precision. If you are able to show newer patterns in the data within a reasonable range of error percentage, you have achieved your task.

- **Provide actionable insights.** Insight generation is critical for innovation. However, just generating insights is not enough; you should strive to provide means to make the insights actionable.

- **Collaboration and reusability.** While solving a particular problem, a data scientist may be following a different set of processes, tools, and approaches. The data sets you use, the models you develop, and the visualizations you create, all need to be documented so that others can understand what you did and how you did it. The goal is to get the capabilities into the hands of analysts in business units, allowing them to create analytic reusable workflows following the path you have taken.

A Day in the Life of a Data Scientist

Data science is a multi-disciplinary set of skills bringing together scientific methods, data and software engineering approaches, statistics, and visualization techniques. This section is not an elaborate discussion of these disciplines but meant to bring all of these different skills together to describe various activities a data scientist needs to perform during the course of a day.

Data science, as practiced today, requires proficiency in parallel business domain knowledge, advanced statistics, machine learning, and intensive programming skills like map-reduce computing, petabyte-sized No SQL databases, etc. In addition data science is also about having a mind-set that lends itself to experimentations and to the ability to construct a story line around data. In Figure 9-4 is a schematic diagram outlining the activities a data scientist performs during the course of solving a business problem.

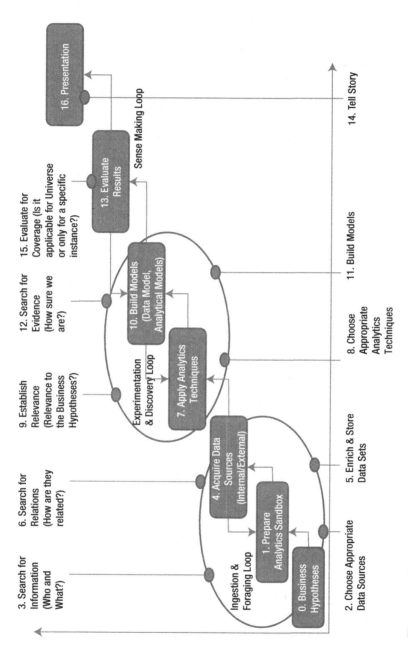

Figure 9-4. Data scientist activities

Thinking about the Problem

Let's start with an example:

The telecom industry, and in particular the communication service providers (CSPs), must have a more in-depth, personalized, or contextual understanding of a subscriber's behaviors and preferences in order to stay competitive.

Some of the business challenges could be:

- Customer churn rates are high, and there is a downward trend in market share capture.

- Predicting bandwidth demands and assuring constant supply is becoming difficult.

- Competitive pressures demanding improved product quality and assurance.

- Developer ecosystems and third-party data providers cannot be ignored.

Translating these business problems to a data scientist set of problem statements would look like the following:

- Predict demand by examining past customer behavior usage and device logs and correlate it with external data sources such as social events.

- Identify churn patterns from support logs, device error logs, and transactional data for proactive customer relationship management.

- Identify and segment customers based on device usage and mobile current logs for better success in promoting new offers, up-selling and cross-selling.

- Gather sensor data from access points and routers to stay ahead of bandwidth fluctuations.

- Analyze social media buzz around events and news and determine its impact on device usage.

- Predict which devices will need to be repaired or replaced by analyzing mobile phone and device logs.

- Run analysis on device error logs for issues such as dropped calls and bad quality of reception.

Data Ingestion and Foraging

Traditionally CSPs have operated with complex disparate silos of data, making data analysis across all the business portfolios extremely challenging. Secondly, as you can see from the problem statements above, you will need a vast range of data and also different types of data to perform your analysis. You will have to pull data from enterprise systems such as billing, customer care, call detail records, and network data to develop a subscriber's behavioral model and to understand a customer's habits such as what delivery format subscriber prefers or their call frequency to off-net users. You will also need external data originating from social networks to determine subscriber's influence within their social circle. You can potentially use the social network data to develop targeted marketing campaigns as well as initiate real time actions toward customers while at the same time notifying network services and IT systems.

Increasing revenue per user is always a CSP's priority, by developing a comprehensive data platform consisting of both internal data and external data, you can develop advanced offer management solutions to help create more innovative and targeted offering and campaigns for up-selling, cross-selling, new customer acquisition, and viral marketing. Effective usage of location-based data can also play a significant role in increasing revenue for mobile services, location-based data enables location based advertisement, which provides new revenue opportunities for wireless carriers.

Profitability, customer churn reduction, and increase in wallet share solely depend on obtaining data that is coherent and current across CSP's entire business portfolios. Making sense of structured and unstructured data to understand the behavior and transaction patterns of customers in real time is critical, as is social network and sentiment analysis. Effective usage of the variety of data sources will help operators take preventive actions so that they can avoid churn or customer dissatisfaction by providing targeted promotions or preemptive service assurance at the moment and in a way that is most relevant to the customer.

Use of traditional metrics, such as Average Revenue Per User (ARPU), Minutes of Use (MOU), Count of Customers, and Churn (i.e., adds, disconnects, transfers, migrations) are useful to analyze the current state of business; however, there is a greater need to analyze data to understand customer experience across fixed, mobile, broadband, and media entertainment services. There is a general understanding that next-generation services encompass seamless communication and media consumption experience for the customer through leveraging high-speed broadband networks, smart devices, and media rich content irrespective of whether the communication network is fixed, wireless, or mobile.

As a consequence of the changes in customer behavior, customers are more likely to churn. CSPs can, however, retain and grow their customer base by investing in next generation technologies and rich media content partnerships. CSPs can also provide seamless customer experience across fixed and mobile networks by means of analytics on customer behavior to provide relevant content and services that are perceived as valuable to them.

To improve service qualities, you will need real-time feeds from network and back office systems. These data will enable you to analyze network performance, service catalogs, and service fulfillment around advanced network functions like real time network planning to improve service performance by correlating subscriber information with network performance.

Experimentation and Discovery

Now that you have an idea about the type of data sources you need to acquire, now it's time to understand what kind of experimentation and discovery you can do with this acquired data. Table 9-1 outlines a high-level view of activities.

Table 9-1. *Use Cases and Data Discovery Activities for a CSP*

Use Case	Description
Up-selling	Identifying optimal targets for a new 4G service launch and triggering usage stimulation through an appealing top-up offer
Cross-selling	Identifying subscribers who seem to travel often and offering them their own personal Wi-Fi device or bundlers with data roaming option, a product that they may not currently have
New customer acquisition	Properly identifying influencers who seem to have many off-net contacts and making them offers that they may spread through word-of-mouth or virally to their off-net family and friends
Multi-SIM prediction	Preventing customers from buying SIM cards from other CSPs by offering them more appealing rates or product bundles
Rotational churn identification	Identifying and preventing mobile subscribers from abusing new handset offerings
Churn location	Identifying and sending more appealing offers or even contacting subscribers located in areas that have a higher churn rate
Dynamic profiling	Analyze incoming data sources like customer care, product/service/device portfolios, cost and billing and network service quality to segment customers by: • Usage – voice, data, SMS usage, times of day • Interests – gaming, music, video, time spent on social media portals • Location based needs/services • Socioeconomic class – prefers the newest high- end devices • Influence in their network – what type of influence they are having within their cluster such as their family, business community, peers • Propensity to churn • Relationship with off-net users (making frequent calls to those using a different provider)

The discovery and experimentation set of activities acts as a major enabler, as it helps navigate through the vast array of data sources to get more in-depth contextual profile of subscribers to understand factors such as customer preferences, usage patterns and predict their future behavior patterns. By applying advanced analytics techniques to this vast array of data you will be able to discover patterns of significance across the data sets and perhaps also provide root cause, predictive and outcome analysis, complex event analysis, and multivariate business activity monitoring intervention opportunities.

As a result of the attractiveness of the content provided by the ISPs and enabled by the freedom of the Internet, customer behavior is expected to transcend a network controlled by the CSP (i.e., Managed Network) to the "unmanaged" Internet (also known as over-the-top). Such a customer behavior can be described by the concepts of time shifting (Prime Time vs. My Time), place shifting ("I decide where I want to watch"), and media mobility or device shifting ("I decide how I want to move and share media between different devices."). Figure 9-5 illustrates a typical customer usage reflecting the points discussed here.

Name: Customer XYZ
Account No.: 10029277
Billing Period: 1-Nov-2012 to 30-Nov-2012

Date	Time (HH:MM:SS)	Event Description	Charge Amount
1-Nov-2012	10:22:30	Data Usage: 4.21 Mb sent; 1.8 Mb received	$9.89
3-Nov-2012	15:13:03	Weather Report: Location – Srinagar	$1.50
4-Nov-2012	8:25:12	Friends and Family Voice Chat (1Hr 15 Mins)	$3.75
6-Nov-2012	9:30:30	MMS Photo Messages	$7.36
8-Nov-2012	18:10:30	GPSNavigation: Location – Kolkata Metro	$5.99
10-Nov-2012	20:30:05	Data Usage: 12.21 Mb sent; 14.8 Mb received	$19.89
11-Nov-2012	21:10:00	Music Album Download: 20.18 Mb	$15.00

Figure 9-5. Illustrative customer usage of CSP services

From the above illustrative data sample, it is evident that critical information, such as presence ("Is the user I want to reach currently on the network?"), identification ("Is the user who he says she/he is?"), and location ("Where is the user?"), can be derived and value-added services can be delivered to the consumer. In addition you can dwell upon additional set of relevant questions:

- How do I charge for a person-to-person, multi-media message?

- How do I determine the characteristics of receiving device and/or network such that I can alter delivery mode and price scheme according to device/network capabilities?

- In a market where calling-party-pays is the norm; who pays for the value-added push services initiated by the service provider's push initiator application?

- How can the growing segment of teenage and student users participate in device-initiated m-Commerce services if they don't have credit or debit card accounts?"

Figure 9-6 illustrates set of activities that a data scientist would perform during the discovery and experimentation phases.

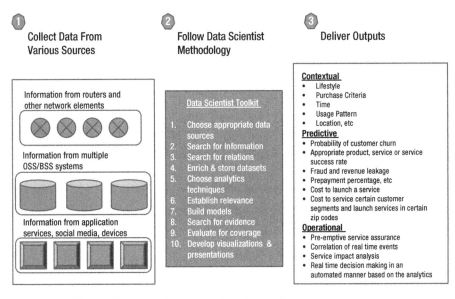

Figure 9-6. *Illustrative data scientist activities for a CSP*

Turning your focus to social network data, you can analyze customer sentiments and identify trends in several areas. Do your customers like you? Do they prefer your competition? Are your marketing messages resonating? If so, why? If not, what are their issues? Sentiment analysis helps organizations understand what customers think about their brand and products. The ability to track customer sentiment gives device and service providers the insight they need to determine where and how to prioritize change.

Knowing whether customers feel positively or negatively about their brand, products, or services gives telecom companies a high-level view into their market acceptance and consumer perception. But what are the issues driving that sentiment? These are the key details service providers and device manufacturers need to know to make needed changes and drive their business forward. Today customers often turn to the Internet when they encounter a problem with a product or service. By tracking issues over time, providers can get a sense of both emerging issues and issues that have gone from a "normal" level of complaint to a serious level. With this insight, the manufacturer or provider can take action to prevent a minor issue from spiraling into a customer service or public relations problem.

The data scientist can use advanced analytics techniques to identify specific product quality issues. For example, the report in Figure 9-7 shows a summary of the top issues identified by customers in social media regarding hand- set quality issues. First shown are the overall general negative issues discussed about the hand-set. Second to these general complaints are issues with the battery life, button keys, the camera not working and issues with the screen display, etc.

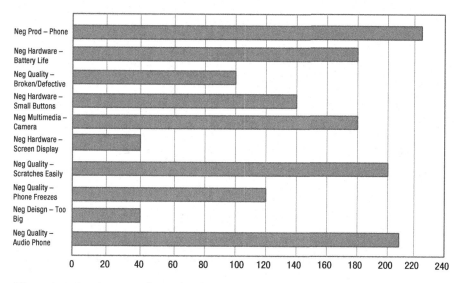

Figure 9-7. *Sentiment analysis related to the quality of the product*

In some cases, the data scientist might even identify issues that could pose a safety threat. The analysis in Figure 9-8 shows the progression of a potentially dangerous issue around a new device that "gets hot" during use. The data scientist has created a time series chart that tracks the issue and the frequency with which the issue was mentioned over time.

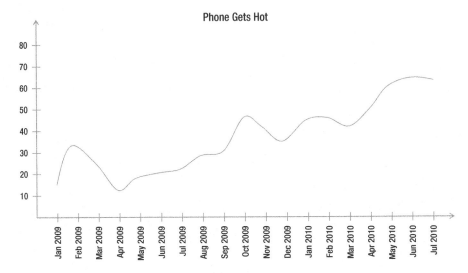

Figure 9-8. *Sentiment analysis: Phone gets hot*

The most challenging part of getting to insight in large volumes of customer conversation data is the great variances in the way people say things. The data scientist can use advanced text analytics solutions and automatically aggregate mentions of churn indicators that are articulated in different ways into a single category. In Table 9-2, churn articulation includes comments such as:

"I am going to switch."
"I am going to move carriers."
"I am going to jump carriers."

Table 9-2. *Sentiment analysis of churn related messages and associated conversation counts*

Message	Count
{Unspecified}:Switch to: Carrier X	100
I:Switch to: Carrier X	250
We:Switch to:Carrier X	220
{Unspecified}:Switch to: Carrier Y	80
I:Move to:Carrier Y	73
I:Switch back to:Carrier Y	287
I:Jump to:Carrier X	83
My Family:Switch to: Carrier Y	34

Evaluation of Results

Deciding which algorithm to apply on the set of data depends on the type of data (interval or categorical, paired vs. unpaired) being analyzed and whether or not the data is normally distributed. Interpretation of the results of the analysis relies on an appreciation and consideration of the null hypothesis, P-values, and the concept of statistical significance.

By constructing a histogram or frequency curve you will be able to understand whether the data follows a normal distribution or not. You can also do box plots to determine if there are outliers in the data sets. Conducting tests like principal component analysis will also help you determine which attribute or set of attributes are influencing the spread of the data and why.

Following are few high-level tips for you to consider:

- Identify the dependent variable. What are you trying to predict?

- Identify the independent variables, or the predictors of the dependent variable.

- Find the statistically significant relationships between independent variables and the dependent variable. The usual standard for statistical significance is less than a 5 percent chance that a relationship this strong would be observed by coincidence, where no real relationship existed. Look for one of the following indicators: p (should be .05 or less), Z score, *significance level*, or the use of *asterisks* (**) to indicate significance at the .05 level or less. In each case, lower numbers are better, since the number is the probability of this relationship being generated by random coincidence.

- Now that you know the statistically significant independent variables, check the direction of the relationship. You are looking for a number that will be called a coefficient, or beta, or β.

  - If you see numbers in parentheses, ignore them. These are usually standard errors, which are used to calculate p. Since you already have p, you don't need them. Look for the number without parentheses: that is the coefficient you want.

  - In most analyses, if this number is positive, then the relationship between the independent variable and dependent variable is direct. Increases in the independent variable increase the value of the dependent variable. If the number is negative, then the relationship is inverse. Increasing the independent variable decreases the value of the dependent variable.

  - In analysis of duration (how long a campaign runs, how long before you see churn indications, how long a product stays as number one in most selling list, etc.), the coefficient often describes an effect on the hazard rate. The hazard rate is the likelihood that some process stops (i.e., product drops from biggest-selling items). In this case, a hazard rate above one means a shorter duration. Similarly, a hazard rate below 1 means a longer duration. The analysis will nearly always tell you whether the coefficients represent effects on the dependent variable or hazard ratios, but you might have to glance at the variable descriptions to see which approach is used.

- Now you know which variables matter and whether each one increases or decreases the significance of dependent variable. Next you need to determine the importance of the variables.

    - How well does the model perform? There is usually some indication of whether knowing all of the independent variables actually helps one predict the value of the dependent variables. There are two main types of information you might see:

        - The chi-square (χ^2) statistic: If this is "big enough" it means the overall model performs better than chance. In the output, look for a significance level printed next to the chi-square statistic. This is just like the significance for the independent variables but applies to the entire combination of independent variables included in the analysis. Remember, this only tells you how likely it is for random chance to have produced these results. It doesn't tell you how much better the model is than flipping a coin.

        - The R-Square (R^2) or Pseudo-R^2 statistic: This is a better measure of how well the statistical model performs. It indicates how much error in guesses about the value of the dependent variable is eliminated when you actually know the values of the independent variables, as opposed to just guessing the average. Example: The average (modal) top-ranking students always get highly paid jobs. Suppose I predict that every top-ranking student gets a high-paying job. I will be right two-thirds of the time. If I use independent variables to predict the dependent variable getting a high-paying job, then an R^2 of .5 would mean I was right about five-sixths of the time (I reduced the number of mistakes by half, or 0.5). Higher numbers are better.

    - Which independent variables are most important? Just because a variable is statistically significant doesn't mean that it has a significant effect on the dependent variable. In order to find out which variables have the most significant effect on the dependent variable, there are two choices:

        - Hope the data scientist included a table with the substantive effect of each variable

- If and only if two variables are measured using the exact same scale (i.e., both are measured in dollars, or both are measured in number of people, etc.) then you can compare their coefficients. The bigger coefficient has a larger effect.

- Repeat the above-mentioned steps for each model summarizing the effects of four or five different combinations of independent variables on the dependent variable. There are usually good reasons for this approach, but it makes it a bit harder to interpret the data. When you see multiple models, look for any independent variables that have significant coefficients (of the same sign, i.e., positive or negative) across all.

- All statistical tests start with the premise of the null hypothesis. This is then tested by calculating the probability that the differences observed between the sample groups are due to chance (the P-value). It is almost always appropriate to conduct statistical analysis of data using two-tailed tests. A one-tailed test is usually inappropriate. It answers a question similar to that of the two-tailed test, but crucially it specifies in advance that we are only interested if the sample mean of one group is greater than the other. If analysis of the data reveals a result opposite to that expected, the difference between the sample means must be attributed to chance, even if this difference is substantial. For example, say you are analyzing the effect of a campaign on product sales; you will collect data before and after the campaign. The data is then analyzed using a paired t-test (as the data are matched pairs of pre- and post-campaign for the products). The data scientist decides to use a one-tailed test, as he is certain that product sales figures must improve after the campaign and discounts the possibility that product sales performance won't score as well after it. Somewhat surprisingly, after the data are analyzed, the mean scores post-campaign are worse than pre-campaign with a P-value of 0.01. The correct statistical interpretation of this result is to attribute the observed difference to random chance. However, it may be indeed be true that the product sales performance went down after the campaign. Perhaps the right target community was not picked up, the messaging in the campaign was confusing, or a competitor's product also got launched at the same time, etc. The data scientist will get wrong inferences if he uses a one-tailed test in this situation—a two-tailed test would have been appropriate.

Presenting the Results

To start with, it is good to adopt the following best practices when you are beginning to develop visualizations to present your findings:

- **Clarity and Context:** This is all about how quickly the user can understand what data the visual is displaying, and how it is displaying it. The visualizations should serve as a means to effectively interpret and explain the underlying data. The visualizations should also be able to establish the context behind the data and highlight the messages you wanted to come forward, such as effectively showing the trends that are connecting the events and explaining the relationships in the data elements

- **Completeness and Connected:** You have prepared your big data set by mashing up data from several sources. Your visualization should seamlessly connect all these different data sources; it should not come across as incoherent. Special care needs to be taken to bring out a complete story, otherwise the insight you have generated will come across as several different observations stitched together to provide a make-believe observation.

- **Focus and Concentration:** In our pursuit to develop glossy and eye-catching visualizations oftentimes we get carried away and put in too many bells and whistles that act as a distraction. The important aspect of visualization is how well the visualization brings certain (sets of) data points to the forefront and helps the viewer focus on them. You should display objects that are vital to the accurate interpretation and contextual understanding of the underlying data: avoid all design aspects that are unconnected to the task of analytic communication. Moving features are good and bad, as they get the attention of users, but too many of these distract the user from other important information.

Conceptualizing an Effective Data Visualization

Conceptualizing data visualization is a complex process. It requires in-depth knowledge of the business context behind the data, lots of creativity, and deep technical knowledge of how to go about implementing it. But another aspect that is often overlooked is user experience: an understanding of how the end user will interact with the visualization. A poor understanding of user experience leads to poorly designed visualizations, which slows down decision making and defeats the very purpose of the visualization. In subsequent sections below, you will understand basic understanding of how end users interact with data visualization and how you can conceptualize your data visualizations better.

What is Business Data Visualization?

A business data visualization is a multi-layered data representation that allows users to understand, interpret, monitor, analyze, and manage their business processes more effectively. Data drawn from various business-related events are represented in a lucid and simple way often bringing in insights from disparate sources and collectively then rendering on highly visually appealing environment.

Components of a Business Data Visualization

Business data visualizations ideally should allow visualization of data at three levels: summarized view, multi-dimensional view, and detailed view (Figure 9-9).

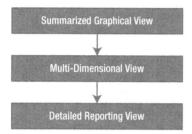

Figure 9-9. *Hierarchical view of data*

It is important to understand that users don't want to see all the data at one place, hence the hierarchical view of the data is always preferred.

The summarized graphical view of the visualization allows a user to get a high-level view of the business process or business event of interest. In case there is a specific view highlighted (usually represented through colors such as red, amber, and green), the user then drills down to the next level of detail, which is the multi-dimensional layer. The multi-dimensional layer brings in several connected areas of data and provides a more detailed view of the event of interest, thereby allowing the user to get a better perspective. Many times, there is still a need to go to past the granular level of data to the transactional level: the user drills farther down to the detailed view layer. The detailed view layer provides individual transactions from where the user can easily articulate root causes.

While building data visualization, data scientists should establish well-defined levels of abstraction with respect to data. This not only makes the visualization user-friendly but also makes it deployable at all levels of organizational hierarchy.

The data design specifications of each data presentation layer are:

> **Summarized View:** This is the topmost layer of the visualization, and it should display essential KPIs graphically (i.e., using graphs or gauges). The layer must also have an in-built control mechanism that triggers an alert when a KPI exceeds or drops below the normal value range.

Multi-dimensional View: This layer should supplement the metrics displayed in the top layer with additional data. Analytical tools must be built into this layer in order to allow users to perform computational analysis on data.

Detailed View: This layer should facilitate users to view reports pertaining to individual transactions (e.g., invoices, shipments, etc.).

Over the past several years, infographics have caught the fancy of business users. Infographics are visual representations of information, implying that sets of data are displayed in a unique way that can be seen, rather than read. These visualizations should not be left up to interpretation; it should instead be designed in a way that provides a universal conclusion for all viewers.

Here are some useful data visualization approaches you should follow when you start designing your data visualizations.

Line Graphs

A line graph shows the relationship between two variables. They are most often used to show changes or trends over time. Line graphs, if shown as stacking lines, display comparison of multiple items over the same time period. Figure 9-10 displays several aspects of growth associated with an IT company over a continuous span of time.

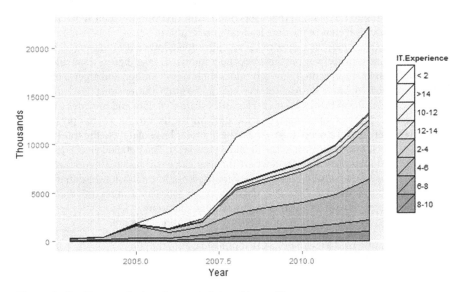

Figure 9-10. *Line graph showing growth trend in an IT company*

Figure 9-11 provides a trend of one variable compared to a moving average (upper bound and lower bound), in this case the salary paid to employees are tracked against an industry average over a continuous time frame.

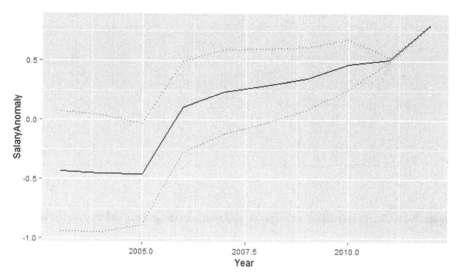

Figure 9-11. *Line graph depicting comparison of salaries paid to the employees against the industry standard over the years*

■ **Note** You should be using line graphs when the change in a variable or variables clearly needs to be displayed and/or when trending or rate-of-change information is something you would like to highlight to the users.

Bar Graphs

Bar graphs are most commonly used visualization technique. These are used for comparing the values of different categories. Values of a category are represented using the bars with the length or height of each bar representing the quantity.

Bar graphs are very effective when the values are distinct enough and the differences in the bars can be easily detected. When the values (bars) are very close together or there are large numbers of values (bars) that need to be displayed, bar graphs become clumsy and it becomes difficult to compare the bars to each other.

Figure 9-12 provides a trend of anomalies in salary given by an IT company over the years, interesting to note the positive and negative variances compared to an industry average.

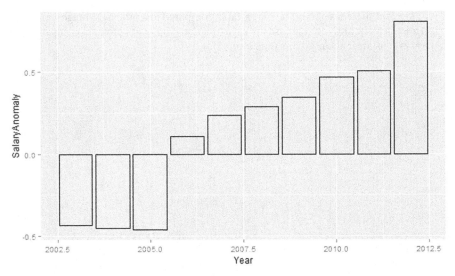

Figure 9-12. *Bar graph depicting negative and positive variances of salary over the years*

Another form of a bar graph is called the "progressive bar chart," or "waterfall chart." A waterfall chart shows how the initial value of a variable increases or decreases during a series of operations or transactions. The first bar begins at the initial value, and each subsequent bar begins where the previous bar ends. The length and direction of a bar indicates the impact (positive or negative, for example) of the operation or transaction. The resulting graph is a waterfall that shows how the transactions or operations lead to the final value of the variable.

Figure 9-13 shows the impact of several contributing factors to the cash flow and final balance of an IT company.

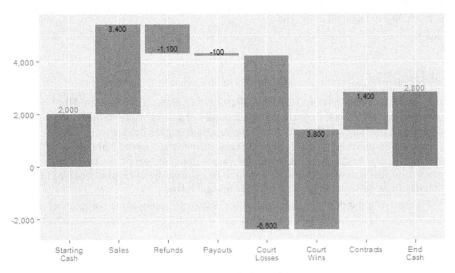

Figure 9-13. *Waterfall graph depicting the cash flow of a company*

274

Scatter Plots

A scatter plot is useful for examining the relationship or correlations between X and Y variables. Variables are said to be correlated if they have a dependency on or are somehow influenced by each other. For example, for an IT company, the project team size is often related to team proficiency index: the relationship that exists might be that as project team size increases the team proficiency index decreases (a negative correlation). A scatter plot is a good way to visualize these relationships in data.

Once you have plotted all of the data points using a scatter plot, you will be able to visually determine whether data points are related. Scatter plots can help you gain a sense of how spread out the data might be or how closely related the data points are, as well as quickly identify patterns present in the distribution of the data.

Figure 9-14 shows a scatter plot taking into account two variables, "project team size" and "team proficiency index." It also shows the impact of these two variables with respect to the "client satisfaction score." You can see that the smaller the "project team size" the higher the "team proficiency index" leading to higher "client satisfaction score."

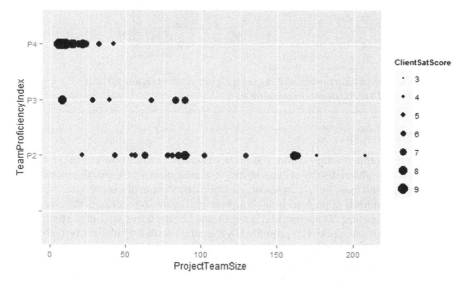

Figure 9-14. *Scatter plot graph depicting the relationships between various groups of data points*

Figure 9-15 shows a correlation matrix taking several variables into account. You can draw many interesting inferences from the graph; for example, proficiency ratings of an individual are associated with the number of SME reviews and white papers published, less time on the bench (idle time). Similarly, the performance rating of an individual is associated with the number of SME reviews, white papers published, and less time on the bench. Another interesting observation: people who've published white papers and who've contributed to assets creation and SME reviews get excellence awards.

275

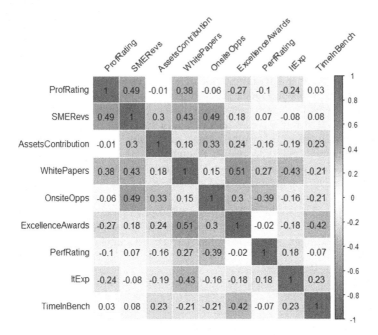

Figure 9-15. *Correlation matrix plotting several contribution factors for SME (Subject Matter Experts) in an IT company*

Box Plots

Box plots are another example of how the volume of data can affect how a visual is shown. A box plot is a graphical display of five statistics (the minimum, lower quartile, median, upper quartile, and maximum) that summarizes the distribution of a set of data.

The lower quartile (25th percentile) is represented by the lower edge of the box, and the upper quartile (75th percentile) is represented by the upper edge of the box. The median (50th percentile) is represented by a central line that divides the box into sections.

Often, box plots are used to understand the outliers in the data. Generally speaking, the number of outliers in the data can be represented by 1 to 5 percent of the data. With traditionally sized data sets, viewing 1 to 5 percent of the data is not necessarily hard to do. However, when you are working with massive amounts of data, viewing 1 to 5 percent of the data is rather challenging.

Figure 9-16 shows a box plot that while most of the data points related to "team proficiency index" and "project team size" are consistent, there is an outlier: a project team size of more than forty resources showing the highest level of team proficiency index.

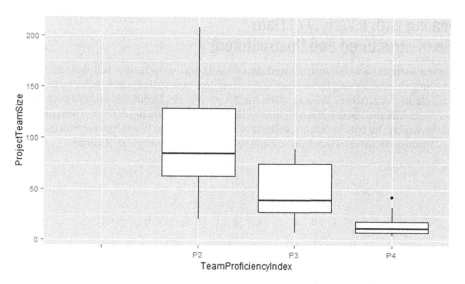

Figure 9-16. *Box plot graph depicting the outlier in data distribution: Project team size vs. team proficiency index*

Figure 9-17 shows another interesting scenario where most of the data points related to "team proficiency index" and "client satisfaction" score are consistent, but there are two instances where even if project teams have high "team proficiency index" the client satisfaction index is not at the expected level.

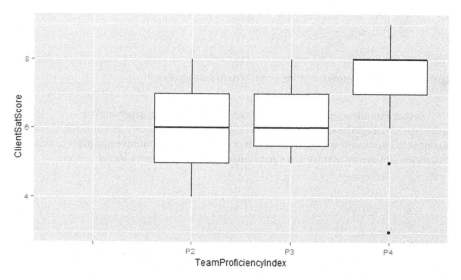

Figure 9-17. *Box plot graph depicting the outlier in data distribution: Client satisfaction score vs. team proficiency index*

Dealing with a Variety of Data (Semi-structured and Unstructured)

The semi-structured and unstructured data do not necessarily have a well-defined structure; thus, analyzing these types of data requires new visualization techniques. A word cloud visual (where the size of the word represents its frequency within a body of text) is an effective visualization technique that can be used on unstructured data as a way to display the concentration of words of interest (high- or low-frequency words). Figure 9-18 shows a word cloud depicting most used packages in R library.

Figure 9-18. Word cloud depicting the most used packages in R

Another visualization technique that can be used for semi-structured or unstructured data is the network diagram (Figure 9-19), which can, for example, show the relationship between several groups of subjects. The most common example is to analyze 'leaders and followers' involved in a particular topic in the tweeter social network.

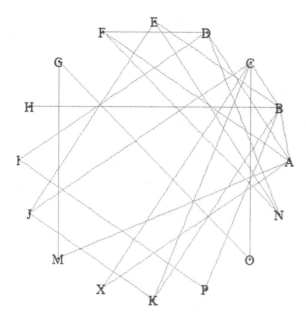

Figure 9-19. Network graph plotting mentors and followers relationship in an IT company

Visualization Velocity

Velocity is all about the speed at which data is coming into the organization. The ability to access and process varying velocities of data quickly is critical. A correlation matrix combines big data and fast response times to quickly identify which variables are related. It also shows how strong the relationships are between variables.

A correlation matrix combines big data and fast response times to quickly identify which variables among the millions or billions are related. It also shows how strong the relationship is between the variables.

Tell a Story

All good stories have a beginning, middle, and end. Data visualizations deserve the same treatment. At the beginning of the visualization, introduce the problem or business hypothesis. From there, back it up with data. Finally, end the visualization with a conclusion.

Visualize the Hook

Every good visualization has a hook or primary take away that establishes the theme. As a designer, you should make this hook the focal point of the design if possible. Placing the hook at either the center or very end of the visualization is usually best, since this will get more attention. Give the most important information the most visual weight so that viewers know what to take away.

With all of the data that goes into the visualization, make sure that the viewer's eye easily flows down the visual; the wrong color palette can be a big barrier to this. Choose a palette that doesn't attack the senses. And consider doing this before you start designing, because it will help you determine how to visualize the various elements.

If picking a color palette is hard for you, stick to the rule of three. Choose three primary colors. Of the three, one should be the background color (usually the lightest of the three), and the other two should break up the sections. If you need to add other colors, use shades of the three main colors. This will keep the palette cohesive and calming rather than jarring.

While these standards are important to consider for most visualization designs, sometimes a new creative idea comes along that breaks all of these rules and still succeeds. Use these rules to guide you into the data visualization realm, but create your own techniques and standards after you've gained some experience.

Here is another example of how to construct a story line using several sources of disparate data.

Tracking Hurricane Sandy!

Hurricane Sandy was a major hurricane that hit the East coast of USA recently and caused extensive damage to the eastern sea board communities. Figure 9-20 shows the path of the hurricane, where it finally headed to landfall and the extent of the impact from the hurricane.

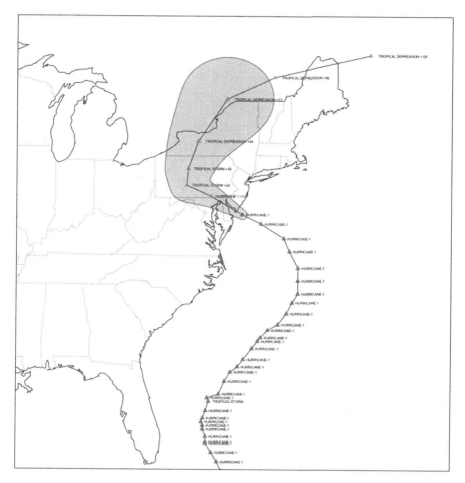

Figure 9-20. *Tracking hurricane Sandy*

The visualization was developed taking real-time time meteorological data to track the path of the hurricane in real-time.

From the visualization above, you can see Atlantic City and Philadelphia were directly in the path. Using further meteorological data, analysis of the local weather conditions during the passage of Hurricane Sandy is shown in Figure 9-21. Atlantic city experienced a faster and larger drop in barometric pressure during the same time period as compared to Philadelphia adding to the obvious ocean front sea wall effects and causing a greater scale of destruction in Atlantic City compared to Philadelphia (refer to Figure 9-22 for a comparison of wind speeds over the same period).

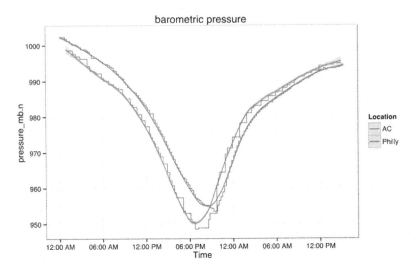

Figure 9-21. *Barometric pressure across two cities on the path of hurricane Sandy*

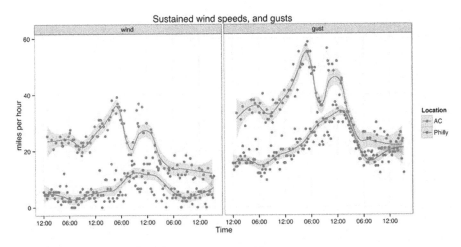

Figure 9-22. *Wind speeds and gusts across two cities*

Source for Figure 9-22: Analysis of a Philadelphia Weather Station Data during Hurricane Sandy. http://rpubs.com/JoFrhwld/sandy

End Points

As discussed above, data scientists use a combination of skills to investigate big data looking for ways to solve business hypotheses and also to create possible new business opportunities. The BI user and data scientist are at the two opposite ends of the spectrum: a BI user analyzes existing business situations and operations, whereas a data scientist investigates and looks for new possibilities.

Data scientists use the discovery type of approach, bringing in various types of data into an investigative data store for experimentation. Depending on the enterprise data platform architectures, this data store could be a separate sandbox; or it could be an integrated environment with the EDW and other enterprise data repositories. Figure 9-23 illustrates the data discovery platform for a data scientist.

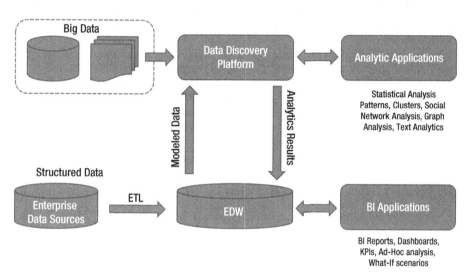

Figure 9-23. *Data discovery platform for a data scientist*

Most of the concepts behind data scientist's activity are not new; for example, statisticians and analytics resources have been building predictive models for many years for risk analysis, fraud detection, and so forth. What is new about data science is that the use of big data and associated data analysis technologies enables a broader set of business solutions to be addressed.

It is evident that the data scientist should possess capabilities to juggle around three specific areas:

- The BUSINESS context behind the data
- The ANALYTICS capabilities to apply on the data
- The DATA itself

Most of the time it's easy to assess the depth of the data scientists analytics/ algorithmic knowledge and the depth of his/her understanding on handling high- velocity data and unstructured data elements. But one area of weakness is the business dimension.

So how do you decide whether a data scientist the desired level of business acumen? Below are few best practices that will help you assess and decode the business domain knowledge of a data scientist.

Test-1: "Resonant Story Telling" Test:

Can the data scientist narrate a compelling and resonant story from the data patterns?

We human beings are naturally wired to listen to stories than to read numbers. Hence it is extremely important for the data scientist to craft a story from the data. Let's look at an example: you are analyzing mobile app funnel drop for an online travel agency and found out that the mobile user who was getting dropped was a twenty-something, last-minute booker traveling between metros and trying to complete the transaction from a Samsung mobile using Android OS and the friction point was the payment gateway.

Can the data scientist translate numbers into stories? This is a very important tool to connect to businesses with. Otherwise, a data scientist runs the risk of getting stuck in the world of math and unable to make the connection!

Test-2: The "String of Pearls" Test:

Can the data scientist connect the dots and form a "necklace" from the pearls of insights discovered from seemingly unconnected data?

It's very important for a data scientist to triangulate from key insights. For example, say you are working on a telecom security use case, and you were able to spot a correlation between multiple failed login attempts and a successful patch download event and a surge in network traffic, which was a result of the security hole in the patch that was downloaded. Can the data scientist connect these seemingly unconnected events and form a "necklace" from the pearls of insights discovered from cryptic log file data points?

Test-3: "Needle Movement" Test:

Which are the best "impact zones" for use cases which are "ripe" for big data?

One of the biggest risks in a big data project is using data and multiple analysis paths to solve the right problem. There are many use cases a data scientist can curate. How do you identify the use cases that are worthwhile from the use cases which have marginal impact?

Big data use cases can be segmented into two categories: those that move the needle incrementally versus those that disrupt. It is very important to keep this distinction in mind. For example, say you are working with an e-commerce company and your area of analysis is how to decrease the percentage of shopping cart abandonment. You have several approaches available: you can decide to provide more discounts for returning customers, establish a price comparison interface for the same products across a few well- known e-commerce sites, etc. However, let's say that while analyzing the data, you realized that a large percentage of the shopping carts are abandoned at the final stages due to issues in the payment gateway; armed with this insight you recommended enhancement of the payment processing application, and the resulting effect was huge upswing in revenues!

Can the data scientist uncover business themes where a use case can unlock disproportionate revenue-making potential for the organization? How would a data scientist go about finding the business themes to move the needle? Which are the best "impact zones" in a business process which are "ripe" for big data?

Test-4: "Sniff The Domain Out" Test:

Can the data scientist "sniff the domain out" by examining analytical outputs and getting the business to put the numbers in context?

Data-driven domain knowledge can reduce the learning curve required to understand domain and is deeper than theoretical knowledge. A data scientist can glean far more knowledge about the nuances of a business by getting his/her hands dirty on exploratory data analysis (EDA), and eyeballing univariate and bivariate results.

Can the data scientist "sniff the domain out" by examining EDA outputs and getting the business to put the numbers in context?

Test-5: "Actionability" Test:

Is data scientist only generating insights or he/she is also crafting a solution to put the insights into action?

Insights are important, but actionable insights are far more important. You can develop a list of insights, prepare suave-looking presentations with lots of graphs and numbers supporting your findings, and the result could be a feel-good effect; but if you are not able to deliver what actions businesses need, your work is useless! For example, say you are working on a use case to spot the high-value customers who are vulnerable to churn. During your analysis, you were able to find out the factors for churning, you were able to also develop predictive models to identify who is going to churn and when. These are valuable insights, but if you are not delivering a solution to prevent the churn, then all your hard work is wasted.

Besides the insight generation, you also proposed a solution where high-value customers who are vulnerable to churn away are redirected in real time to high touch contact center agents who would call them instantly and offer an instant rebate to woo them back.

As a data scientist, you have a larger role to play in operational actions.

Test-6: "Use Case Curation" Test:

Can you give a raw data set to the data scientist, and can the data scientist curate an interesting possibility from the raw data set?

Carving out new use cases and possibilities from new data pool is both an art and a science. For example, say you were able to use search logs that were typically discarded to decode the travel intent of an online booker: is it a price-sensitive traveler or a value-conscious traveler? Is the traveler an early bird or a last-minute booker? This use case to create behavioral tags from search logs resulted in more intelligent outbound actions.

Give a raw data set to the data scientist and ask him/her to curate an interesting possibility from the raw data set. Where would he/she start? How would he/she formulate the right "catchment" of use cases? What approaches he/she would take?

Test-7: The "North Pole" Test:

Can the data scientist work with business to articulate the "as is" state and the expected "to be" state of the decision-making process after the analytical solution is implemented?

Every big data voyage requires a north pole in terms of measuring success for the engagement. A data scientist must be clear about what constitutes success for the business stakeholders, be it a sandbox setup or a full-fledged production setup of a Hadoop cluster.

Can the data scientist work with business to articulate the "as is" state and the expected "to be" state of the decision-making process after the analytical solution is implemented?

Test-8: The "What do You See" Test:

Can the data scientist convey an easily business understandable set of statements from the complex clustering outputs, keyword frequencies, box plots, and other analytical outputs?

This test is all about interpreting the analytics outputs and presenting this info into an easily understandable format.

The sample analytical model's outputs can be:

- Key word frequencies from text mining

- Scatter plots and box plots measuring behavioral volatility of customer balances

- Bivariate cross tab outputs

- Clusters from a segmentation output

- Confidence scores and lift scores

These are outputs are very well understood by analytics professionals but very difficult to comprehend for a business user.

Can the data scientist construct three to four meaningful English statements from the above sample analytical outputs? Can he/she cross the big chasm from math to a business pattern that can be well understood by businesses?

In summary, these tests are by no way collectively exhaustive or perfect. But the tests serve as a reasonable starting point to identify data-scientist-type resources.

References

Big Data Analytics Architecture: By Neil Raden

http://media.smashingmagazine.com/wp-content/uploads/2011/10/Plane-newest.gif

http://www.submitinfographics.com/full-size-infographics/image-153.jpg

http://viralms.com/images/happy-birthday-twitter.png

http://blog.fluturasolutions.com/2012/12/8-tests-to-decode-business-accumen-of.html

http://datascientits.com/2012/12/19/field-note-what-makes-big-data-big-some-mathematics-behind-its-quantification/

Index